Y.G. Zuo (Ed.)

Proceedings from the ICERP 2016 –
International Conference on Environmental Research and Public Health

Y.G. Zuo (Ed.)

Proceedings from the ICERP 2016 – International Conference on Environmental Research and Public Health

Language Editor: Aoife Cocoran

DE GRUYTER
OPEN

ISBN 978-3-11-055903-3
e-ISBN 978-3-11-055904-0

Library of Congress Cataloging-in-Publication Data
A CIP catalog record for this book has been applied for at the Library of Congress.

Published by De Gruyter Open Ltd, Warsaw/Berlin
Part of Walter de Gruyter GmbH, Berlin/Boston
The book is published with open access at www.degruyter.com.

Language Editor: Aoife Cocoran

www.degruyteropen.com
Cover illustration: © MadamLead / iStock.com

Contents

Preface

The 2016 International Conference on Environmental Research and Public Health (ICERP 2016) was held in Shenzhen, China during October 21st and 22nd, 2016. Shenzhen has been one of the fastest growing cities in the planet since the late 1970s, and now is a manufacture and innovation center, and a delightful place for the international conference, located just to the North of Hong Kong.

The Conference covered a broad range of areas in fundamental environmental science, engineering and related public health. There were two keynote speeches: Professor Yuegang Zuo from the University of Massachusetts Dartmouth talked on environmental estrogenic pollution and its effects on public health, and Prof. Jin-Hyo Boo from Sungkyunkwan University on development of advanced oxide nano-materials for environmental catalysis applications. Three speakers, Prof. Guimiao Lin from Shenzhen University, Prof. Nirajan Dhakal from Spelman College, and Prof. Nan Liu from Institute of Health and Environmental Medicine, presented illuminating talks on “Toxicity studies on quantum dots”, “Climate change and water resources management” and “Novel label-free fluorescence assays for one-step rapid and sensitive detection of Hg^{2+} in environmental water samples”, respectively. The participants, including academic scholars, scientists, engineers, and graduate students, enjoyed the warm discussion on the broad environmental research fields covered by the conference, the opportunity for the exchange of new ideas and application experiences, as well as establishing new business or research relations and seeking for collaborations, and appreciated the efforts made by the Science and Engineering Research Center, Hong Kong for organizing this informative conference in the beautiful Shenzhen.

One hundred and one manuscripts were submitted to the conference. 151 international experts were invited and reviewed papers. 21 papers have been accepted and included in the conference proceedings. The quality of papers has been controlled strictly during the review and selection processes. I would like to thank all reviewers involved for their time and effort in reviewing the manuscripts. Without their commitment it would not be possible to have the “peer-reviewed” proceedings.

Prof. Yuegang Zuo,
Editor of ICERP 2016

Committees

Honor Chair

Prof. Shaocai Yu, Lab for Sources and Causes of Atmospheric Pollution, College of Environmental and Resource Sciences, Zhejiang University, China; Department of Marine, Earth and Atmospheric Sciences, North Carolina State University, USA

Guest Editor

Prof. Vladimir Strezov, Department of Environmental Sciences, Faculty of Science and Engineering, Macquarie University, Australia

Dr. Didier Robert, Institute of Chemistry and Process for Energy, Environment and Health-CNRS, antenna of the Universities of Strasbourg and Lorraine, France

Co-editor

Prof. Yuegang Zuo, Department of Chemistry and Biochemistry, University of Massachusetts Dartmouth, USA

Technical Program Committee

Prof. Vladimir Strezov, Department of Environmental Sciences, Faculty of Science and Engineering, Macquarie University, Australia

Prof. Shaocai Yu, Lab for Sources and Causes of Atmospheric Pollution College of Environmental and Resource Sciences, Zhejiang University, China

Prof. Lin Li, School of Biomedical Engineering, Capital Medical University, China

Prof. Yuanshu Jing, Nanjing University of Information Science & Technology, China

Prof. Zhengjun Liu, Department of Automatic Test and Control, Harbin Institute of Technology, China

Prof. Yinhai Lang, College of Environmental Science and Engineering, Ocean University of China, China

Prof. Fei Qi, College of Environmental Science and Engineering, Beijing Forestry University, China

Prof. Wenjun Zhang, School of Life Sciences, Sun Yat-sen University, China

Prof. Hongzhi Wang, Harbin Institute of Technology, China

Prof. Jianzhang Fang, South China Normal University, China

Prof. Janet Yat-Man Tang, National Research Centre for Environmental Toxicology (Entox), University of Queensland, Australia

Prof. Zhongyu Li, School of Petrochemical Engineering, Changzhou University, China

Prof. Xiuhan Li, School of Electronic and Information Engineering, Beijing Jiaotong University, China

Prof. Wen-Tsai Sung, Department of Electrical Engineering, National Chin-Yi University of Technology, Taiwan
Prof. Yudong Lu, Environmental Science and Engineering, Chang'an University, China
Prof. Hong Chen, Department of Resources and Environment, Southwest University, China
Prof. Chenyin Tung, Department of Health Promotion and Health Education, National Taiwan Normal University, Taiwan
Prof. Weihwa Chiang, Department of Architecture, National Taiwan University Science and Technology, Taiwan
Prof. Jin-Hyo Boo, Department of Chemistry, Sungkyunkwan University, Korea
Prof. Dingchin Chou, Department of Architecture, National Taipei University of Technology, Taiwan
Prof. Kambiz Akbari Noghabi, National Institute of Genetic Engineering and Biotechnology, Tehran-Iran
Prof. Tomohisa Hirobe, Keio University and Chiba Institute of Technology-Chiba, Japan
Prof. Amir Movafegh, Head Radiological Protection Research Dept., Nuclear Science and Technology Research Institute, Teheran-Iran
Prof. Dinghui Zou, Environmental and Energy Engineering, South China University of Technology, China
Prof. Leslie Sue Lieberman, Department of Medical Anthropology University of Central Florida Orlando, USA
Prof. Md. Abul Hossen, Department of Social Work, Jagannath University, Bangladesh
Prof. Thomas C. Voice, Michigan State University, USA
Prof. Andrew Kusiak, Department of Mechanical and Industrial Engineering, The University of Iowa, USA
Prof. Chuan-rong Zhang, Department of Geography Center for Environmental Sciences and Engineering (CESE), University of Connecticut, USA
Prof. Matt Crosston, Bellevue University, USA
Prof. Jo Ann Oravec, University of Wisconsin at Whitewate, USA
Prof. Yuegang Zuo, Department of Chemistry and Biochemistry, University of Massachusetts Dartmouth, USA
Prof. Aiqing Guo, Hebei GEO University, China
Prof. Feng Wu, School of Resource and Environmental Sciences, Wuhan University, China
Prof. Guiling Zhang, College of Chemistry and Chemical Engineering, Ocean University of China
Dr. Hong Yang, Assistant prof, School of Public Health, Department of Occupational and Environmental Health, Southeast University, China
Dr. Geoffrey Jalleh, Faculty of Health Sciences, Curtin University, Australia
Dr. Zhiyun Jiang, College of Resource Science & Technology, Beijing Normal University, China

Dr. Jiangyong Hu, Dept of Civil and Environmental Engineering, National University of Singapore, Singapore

Dr. Khair Al-deen Bsisu, Faculty of Engineering and Technology: Civil Engineering Department, The University of Jordan, Jordan

Dr. Alan Woodley, School of Electrical Engineering and Computer Science, Queensland University of Technology, Australia

Dr. Shayfull Zamree Bin Abd Rahim, School of Manufacturing Engineering, Universiti Malaysia Perlis, Malaysia

Dr. Kunquan Li, Department of Agricultural Machinery Institute, Nanjing Agricultural University, China

Dr. Dayu Kao, Department of Information Management, College of Police Science and Technology, Central Police University, Taiwan

Dr. Che Azurahanim Che Abdullah, Department of Physics, Faculty of Science, Universiti Putra Malaysia, Malaysia

Dr. Asral Bahari Jambek, School of Microelectronic Engineering, Universiti Malaysia Perlis (UniMAP), Malaysia

Dr. Jihua Wang, Department of Industrial Engineering, University of Illinois at Urbana-Champaign, USA

Dr. Aimin Yang, Brown-China Center for Environmental Health Sciences, Brown University, China

Dr. Tianxing Cai, Dan F. Smith Department of Chemical Engineering, Lamar University, China

Mr. Andong Zhan, Department of Computer Science, Malone Center for Engineering in Healthcare, Johns Hopkins University, USA

Dr. Quintin L. Williams Jr, Assistant Professor, Department of Mechanical & Industrial Engineering, University of Illinois at Chicago, USA

Dr. Chi-wai Kan, Associate Professor, Institute of Textiles and Clothing, The Hong Kong Polytechnic University, Hong Kong

Mr. Weimo Liu, Software Engineer/Research Scientist at GraphSQL Inc, USA

Dr. Yujia Song, Department of Environmental Engineering, Changchun University of Science and Technology, Changchun, P.R. China

Dr. Rita Yi Man Li, Director, Sustainable Real Estate Research Center, Hong Kong Shue Yan University, HongKong

Dr. Nor Amani Filzah Binti Mohd Kamil, Department of Water and Environment, Faculty of Civil and Environmental Engineering, 86400 Parit Raja, Batu Pahat, Johor, Malaysia

Dr. Ahmet H. Ertas, Biomedical Engineering Department Engineering Faculty, Karabuk University, Turkey

Dr. MD. Mizanur Rahman, Associate Professor, Department of Electrical and Electronic Engineering University of Information Technology & Sciences (UITS), Bangladesh

Dr. Malka N. Halgamuge, Department of Electrical & Electronic Engineering, School of Engineering, The University of Melbourne, Australia

Prof. Yuegang Zuo

Keynote Speech I

Environmental Estrogenic Pollution and Public Health

Abstract: Estrogenic chemicals, particularly the synthetic estrogenic steroids – ethinylestradiol (EE2) and mestranol (MeEE2) and natural hormone steroids – estrone (E1), estradiol (E2) and estriol (E3), have attracted a great deal of scientific and public attention during recent years due to their occurrence in surface waters and sewage treatment plant effluents and their potential adverse effects on the development and reproduction of fish, wildlife and even human beings. In this presentation, we will focus on our research on the occurrence, sources, and microbial and photochemical degradation of both synthetic and natural estrogenic steroids in fresh and marine aquatic environments and their effects on public health during the past decade. To face analytical challenges for determining trace amounts of estrogenic steroids in natural waters, GC-MS and HPLC analytical methods have been developed. The developed methods were applied to the water samples periodically collected from wastewater treatment plants, lakes, Acushnet River and Buzzards Bay. The interesting compounds were detected in several of water samples in nano- to microgram per liter concentration range, in which they can certainly cause fish feminization and may also contribute to the observed declines in lobster population in Buzzards Bay [Zuo et al., Chemosphere 63 (2006) 1583; Zuo et al. Environ. Sci. Processes & Impacts 15 (2013) 1529]. Microbial and photochemical degradation of E1, E2, E3, EE2 and MeEE2 have been also investigated in seawater as well as in waste, lake and river waters as a comparison. The microbial degradation of synthetic steroid estrogens is extremely slow with a half-life of longer than 70 days in seawater. However, the photodegradation of these compounds is much faster with a half-life of 17 hours for EE2 and 19 hours for MeEE2. Humic and other dissolved organic substances significantly accelerate the sunlight-induced photodegradation of estrogenic steroids. Transition metal Fe(III), nitrate and nitrite can further catalyze the photochemical decomposition of these steroids.

Biography: Yuegang Zuo is currently a Full Professor in analytical and environmental chemistry and Director of Graduate Programs at Department of Chemistry and Biochemistry, University of Massachusetts Dartmouth. He is also a Full Professor in marine sciences and technology at the School of Marine Science and Technology, University of Massachusetts. He received his B.S. degree in chemistry from Wuhan University in 1982, his M.S. degree in environmental chemistry from the Research Center for Eco-Environmental Sciences, Chinese Academy of Sciences, in 1984, and his Ph.D. in environmental science from Swiss Federal Institute of Technology Zurich in 1992. Most of his recent research has focused on separation, identification and quantification of endocrine disrupting pollutants and phenolic antioxidants in plants and seafood as well as in the related environments. He examined their occurrence, sources, distribution, transportation, bioeffecs and fate in the biochemsphere. He has published over 70 peer-reviewed papers in prestigious international scientific journals such as Science, and Environmental Science and Technology.

Prof. Jin-Hyo Boo

Keynote Speech II

Development of Advanced Oxide Nano-materials for Environmental Catalysis Applications

Abstract: Today, a major issue about water pollution are the residual dyes from different sources (e.g., textile industries, paper and pulp industries, dye and dye intermediates industries, pharmaceutical industries, tannery and craft bleaching industries, etc.), and a wide variety of persistent organic pollutants have been introduced into our natural water resources or wastewater treatment systems. In fact, it is highly toxic and hazardous to the living organism; thus, the removal of these organic contaminants prior to discharge into the environment is essential. Varieties of techniques have been employed to degrade those organic contaminants and advanced heterogeneous photocatalysts. Involving titanium dioxide (TiO_2) appears to be one of the most promising technologies.

In this study, we synthesized the ZnO and TiO_2 nanoparticles for environmental applications such as water or air purification, utilizing photocatalysts with/without surface modification. Under either UV or Visible light conditions, we investigated the photocatalytic activity tests of the ZnO nanoparticles as compared with both commercial TiO_2 powder (Degussa P-25) and S and Zn-doped TiO_2 nanoparticles. When the ZnO nanoparticles irradiated by UV light, ZnO nanoparticles have less characteristic of catalyzed reduction and oxidation (redox) reactions than those of pure TiO_2 powder and S-doped TiO_2 nanoparticles in presence of O_2/air/water and degraded harmful organic materials. However, when we made an experiment on modifying the surface of ZnO nanoparticles by O_2 plasma surface treatment, the photocatalytic activity of ZnO nanoparticles was increased as much as 60% by O_2. Also, the S-doped TiO_2 nanoparticles and O_2 plasma surface treated ZnO nanoparticles showed relatively high photo-degradation phenomena even under Visible light condition.

An enhancement effect of photocatalytic efficiency by mixing TiO_2 particle and a small amount of Zn particle was also observed with targeted organic pollutant solutions and various amounts of Zn particle added. We analyzed degradation of organic materials as UV-vis spectrophotometer and got a result that a photocatalytic effect is increased with increasing amount of Zn particle.

Biography: Jin-Hyo Boo is currently a professor of Sungkyunkwan University. He received his B.S. degree in chemistry from Sungkyunkwan University in 1985, his M.S. degree in physical chemistry from Sungkyunkwan University, in 1987, and his Ph.D. in Surface & Materials Chemistry from Sungkyunkwan University in 1992.

Most of his recent research has focused on functional materials synthesis, surface & plasma chemistry, nano & semiconductor chemistry, chemical sensor & bio-sensor applications and energy & environmental applications.

Li Ren*, Xinyi Xiang, Weilin Liu

Water Resource Sustainable Utilization Evaluation Index System in Jiangsu Coast Reclamation Region

Abstract: As a comprehensive system, there are many subsystems such as water resource subsystem, social subsystem, economic subsystem and ecological subsystem in water resource sustainable utilization system. In this paper, an evaluation system including three levels is set up according to the metric demands of sustainable water resource utilization in Jiangsu coast reclamation region, namely the target level, the rule level, and the index level. Considering the large number of the indexes, the analytic hierarchy process is used to determine the weights of all these subsystems in the total goal of water sustainable utilization. By analyzing these weights, the attributes of water resource itself is found to be the most important aspect for the evaluation of sustainable utilization in Jiangsu coast reclamation region, and the second important aspect is the situation of the eco-environment.

Keywords: water resource, sustainable utilization, index system, index weight, Jiangsu coast reclamation region, analytic hierarchy process.

1 Introduction

Sustainable development is the overall strategy of social economic development in all countries and regions around the world, in which sustainable utilization of water resource is a new mode of water resource utilization. With the features of complexity, universality, dynamics, and regionalism, it is the most appropriate utilization of water resource integration of exploitation and utilization, protection and management [1]. An important basis of water resource sustainable management is the scientific evaluation of the environment and economic results of water resource exploitation and utilization, which can be used to find out the existing problems [2,3]. To evaluate water resource exploitation and utilization effectively, the first problem that needs to be solved is the establishment and construction of an evaluation index system. This paper researches and sets up an comprehensive evaluation index system to reflect and measure the water resource sustainable utilization in Jiangsu coast reclamation

***Corresponding author: Li Ren,** State Key Laboratory of Hydrology-Water Resources and Hydraulic Engineering, Hohai University, Naijing, 210098, China, E-mail: renli@hhu.edu.cn
Xinyi Xiang, School of geographical sciences, Southwest University, Chongqing, 400715, China, ruoqing@swu.edu.cn
Weilin Liu, Jiangxi Provincial Key Laboratory of Hydrology-Water Resources and Water Environment, Nanchang Institute of Technology, Nanchang, Jiangxi, 330099, China, weilin_liu@nit.edu.cn

region, and decides the weight based on the contents of water resource and its exploitation and utilization by analyzing the relationship between water resource and ecological environment, economy, as well as the society. It provides a scientific basis for water resource sustainable exploitation and utilization in Jiangsu coast reclamation region so that it guarantees the sustainable development of the economic society in Jiangsu coast reclamation region.

2 Concepts of Water Resource Sustainable Utilization

Water resource sustainable utilization refers to the whole process to support the harmonious development of population, resource, environment and economy, and to satisfy the water demands of generational and intergenerational population, maintaining the sustainability of water and integrity of the ecological system [4]. Therefore, water resource sustainable utilization approach is different from the traditional water resource exploitation and utilization in nature. It considers ecological environment values and social values more than only focusing on basic economic values. It's conducted according to the strategy of harmonious development of population, resource, environment and economy. With a clearly defined objective, it embodies the equity principle for human kind to share the environment, resource and economy, and social benefits. We should follow the idea of "unity, harmony, optimization and cycle", use the systematic method as well as high and new technology in implementation to realize equitable and efficient development [5].

3 Evaluation Index System of Water Resource Sustainable Utilization in Jiangsu Coast Reclamation Region

3.1 The Principles to Set up Evaluation Index System

The above theory has clearly stated that the evaluation index system of water resource sustainable utilization should display not only the water resource features, its exploitation, utilization and management, which constitute the water resource system development level, but also the harmonious development level of water resource system with social system, economic system, and ecological environment system [6]. Thus we have to follow the subsequent principles when we select indexes to set up the index system [7]:

1. *The index should have objective scientificity*: The index should objectively reflect the internal relationship of water resource system; provide clear definition, standard measuring methods, and normative calculating methods; and fully reflect the major features and development of water resource exploitation & utilization in the water basin.

2. *The index should have generality*: We should try to select the representative comprehensive index from those that fully reflect the sustainable utilization features of the water resource.
3. *The index should have systematic hierarchy*: It's requested to set up a systematic and well-organized index system with clear hierarchy to express the complicated problem in a compact, clear and hierarchical index system.
4. *The index should have operability:* The index should be simple, clear, and easy to be obtained and operated, with high comparability.

3.2 Hierarchy Structure of the Index System

The evaluation index system of water resource sustainable utilization can be divided into three levels of target, rule and index based on the measuring demands of water resource sustainable utilization and the construction principles of the evaluation index, referring to the construction methods of sustainable development index system, cyclic economy development index system and other systems, according to the practical situation of Jiangsu coast reclamation region. Reflected by the rule level, the target level has a single target. Composed of three subsystems, the rule level is reflected by the index level, which comprises several quantified indexes to completely show the major factors that influence the water resource sustainable utilization in Jiangsu coast reclamation region. See Table 1 for the composition.

Table 1: Evaluation indexes for water resources sustainable utilization in Jiangsu coast reclamation region

Rule level B	Index level C	Unit	Calculation formula
Social subsystem B_1	Population density (C_1+)	Person/km^2	Total population/region area
	Cultivated land per capita (C_2)	Ha/person	Cultivated land area/total population
	Urbanization ratio (C_3)	%	Urban population/total population
	Fiscal revenue (C_4)	Per ten thousand Yuan	See the statistics yearbook
Economic subsystem B_2	GDP per capita (C_5)	Per ten thousand Yuan	GDP/total population
	GDP per unit water (C_6)	Yuan/m^3	GDP/total water resource amount
	Water consumption per unit of GDP (C_7)	m^3/ten thousand Yuan	Water consumption/GDP
	Water consumption per ten thousand Yuan GDP (C_8)	M^3	Production water consumption/GDP

continued **Table 1:**

Rule level B	Index level C	Unit	Calculation formula
Economic subsystem B_2	Irrigation water comprehensive utilization coefficient (C_9)	%	Field water consumption/ water withdrawal at water intake
	Water consumption per ten thousand value-added of industry (C_{10})	m^3/ten thousand Yuan	Industrial water consumption/ value-added of industry
Water resource subsystem B_3	Water resource per capita (C_{11})	m^3/person	Local water resource amount/ total population
	Water production modulus (C_{12})	Per ten thousand m^3/ km^2	Regional water resource amount/land area
	Surface water exploitation and utilization ratio (C_{13})	%	Annual surface water supply amount×100%/local surface water resource amount
	Ground water exploitation and utilization ratio (C_{14})	%	Annual ground water exploitation amount×100%/ allowable ground water exploitation amount
	Water supply per capita (C_{15})	m^3/person	The ratio between water supply and total population
	Ground water supply percentage (C_{16})	%	Ground water supply amount/ available water supply
	Water supply modulus (C_{17})	Ten thousand m^3/km^2	Water supply amount/water supply area
	Water resource exploitation rate (C_{18})	%	Exploitation amount/total water resource amount
Ecological subsystem B_4	Dilution ratio (C_{19})	%	Sewage discharge amount/ surface runoff
	Sewage treatment ratio (C_{20})	%	Sewage polluting input/total sewage discharge amount
	Ecological environment water consumption ratio (C_{21})	%	Ecological environment water consumption amount/total water consumption amount

1. *Target level A:* The target of water resource sustainable utilization evaluation in Jiangsu coast reclamation region is to give a comprehensive evaluation of the comprehensive water resource utilization level in coast reclamation region and to provide decision basis for the sustainable development in the region.

2. *Rule level B:* Rule level is composed of four subsystems: social subsystem, economic subsystem, water resource subsystem, and ecological subsystem. It shows the influence of each subsystem in the index system on the sustainable utilization comprehensive index. This is the core subsystem and main body. The social and economic subsystems refer to the human kind and the surroundings, as well as the industry structure and economic benefits.
3. *Index level C:* It's a group of basic indexes that describe the status of sustainable development. The indexes which are frequently used in comprehensive evaluation research of the current water resource sustainable utilization are selected to show the universality of the index system. In addition, we also pay attention to the feasibility and comparability of the indexes, trying the best to use the existing statistics data.

4 Determine the Evaluation Index Weight

4.1 Analytical Method

Analytic Hierarchy Process (AHP) is a flexible and simple multi-criteria decision making method proposed by an American scholar, Professor T. L. Saaty. At first, to decompose a complicated system into several levels. The specialists evaluate the importance of the indexes in each level and then determine the weight coefficient of each subsystem in the total goal by calculating the contribution degree (weight) of the lower layer indexes to the upper layer indexes. This paper uses AHP to determine the weight coefficient of social, economic, water resource and ecological subsystems in water resource sustainable utilization comprehensive evaluation.

4.2 Figure Captions

1. *Establish the Judgment Matrix:* Decide the relative importance between every two elements in the lower layer according to a certain element in the upper layer, quantize them, and constitute a matrix form, which is a judgment matrix.

$$A = \begin{bmatrix} a_{11} & a_{12} & \cdots & a_{1n} \\ a_{12} & a_{22} & \cdots & a_{2n} \\ \vdots & \vdots & \vdots & \vdots \\ a_{n1} & a_{n2} & \cdots & a_{nn} \end{bmatrix} \tag{1}$$

2. *Evaluate the Geometric Mean of Elements of Each Row:*

$$b_i = \left(\prod_{a_{ij}}^{m} a_{ij} \right)^{1/m}, \quad i = 1, 2, \cdots, m \tag{2}$$

3. *Calculate the Weight Coefficient:*

$$\omega_j = b_j / \sum_{R=1}^{m} b_R, \quad j = 1, 2, \cdots, m \tag{3}$$

4. *Evaluate the Maximum Feature Root of Judgment Matrix:*

$$\lambda_{\max} = \frac{1}{m}\sum_{i=1}^{m}\left(\sum_{j=1}^{m} a_{ij}\omega_j / \omega_j\right) \tag{4}$$

5. *Consistency Check:*

$$\begin{aligned} CR &= CI / RI \\ CI &= 1/m - 1 \times (\lambda_{\max} - m) \end{aligned} \tag{5}$$

In the above formula, CR is the random consistency ratio of the judgment matrix; CI is the general consistency index of the judgment matrix; RI is the average random consistency index (referring to the table) of the judgment matrix; m is the order of the judgment matrix. $CR < 0.1$ indicates that the constituted judgment matrix meets the requirements well.

Table 2: The weights of evaluation indexes for water resources sustainable utilization in Jiangsu coast reclamation region

Target level	Rule level and its weight	Index level	Weight in target level
Water resource exploitation and utilization evaluation in Jiangsu coast reclamation region	Social subsystem B_1 (0.0917)	Population density (C_1)	0.0182
		Cultivated land per capita (C_2)	0.0198
		Urbanization (C_3)	0.0095
		Fiscal revenue (C_4)	0.0441
	Economic subsystem B_2 (0.2696)	GDP per capita (C_5)	0.0297
		GDP per unit water (C_6)	0.0964
		Water consumption per unit of GDP (C_7)	0.0188
		Water consumption per ten thousand Yuan GDP (C_8)	0.0118
		Irrigation water comprehensive utilization coefficient (C_9)	0.0687
		Water consumption per ten thousand Yuan value-added of industry (C_{10})	0.0444
	Water resource subsystem B_3 (0.3889)	Water resource amount per capita (C_{11})	0.0125
		Water production modulus (C_{12})	0.0640
		Surface water exploitation and utilization ratio (C_{13})	0.0819
		Ground water exploitation and utilization ratio (C_{14})	0.0193
		Water supply amount per capita (C_{15})	0.0250
		Ground water supply percentage (C_{16})	0.0176
		Water supply modulus (C_{17})	0.1078
		Water resource exploitation ratio (C_{18})	0.0609
	Ecological environment subsystem B_4 (0.2497)	Dilution ratio (C_{19})	0.1843
		Sewage treatment ratio (C_{20})	0.0419
		Ecological environment water consumption ratio (C_{21})	0.0236

As shown in Table 2, the weight of 4 indexes in rule level and 21 indexes in index level can be calculated based on the above formula. The consistency check should be implemented as required to eliminate the error caused by subjective judgment. The results of weight consistency check are: 0.0820, 0.0047, 0.0228, 0.0734, 0.0123, which are all less than 0.1. It shows that the weight calculation results meet the consistency and the judgment matrix works successfully.

5 Determine the Evaluation Index Weight

To evaluate water resource sustainable utilization, we need to conduct a comprehensive research by combining social economic development and ecological environment protection with water resource status. Therefore the paper sets up a hierarchical evaluation index system that reflects the relationship between water resource, society, economy, and ecological system in Jiangsu coast reclamation region. The proposed index system is a 3-level hierarchical structure composed of target level, rule level, and index level. The specific 21 indexes in it can provide a complete map of the harmonious relationship between water resource and local natural status as well as human activities. Deciding the index weights based on AHP greatly reduces the subjective biases and makes the index configuration more appropriate. It can be found out according to the analysis on index weight value that the nature of the water itself is the most important aspect to measure water resource sustainable utilization in Jiangsu coast reclamation region. And the influence of the ecological environment on water resource sustainable utilization comes next.

All in all, we select proper index evaluation method to make the evaluation based on the determination of indexes and their weights. We discover problems and make proposals to solve the problems; so that we can guide the practice of water resource exploitation and utilization to improve the sustainable development of the ecological environment and social economy.

Acknowledgement: This research was supported by the Jiangsu Province Natural Science Foundation (BK20130842), the Project supported by the National Natural Science Foundation of China (51409091, 51309130) and the Fundamental Research Funds for the Central Universities (2015B14614).

References

[1] Tang L., Zhang W.B., Research on assessment index system for sustainable utilization of urban water resources[J], J.amam, 2012, 212-213, 569-573.

[2] Gai W.Q., Innovation networks—regional economic development[M]. Beijing: Peking University Press, 2002: 16-30.

[3] Hakansson H. Corporate Technological Development: Cooperation and Networks[M]. London: Routledge, 1987: 120-123.
[4] Feng S.Y., An introduction to the sustainable utilization of water resources and the management[M].Beijing: Science press, 2000: 62-214.
[5] Yang J.J., Zhou J., An integrated model for simulating and evaluating water resources sustainability at laoshan region[J]. J.amam, 2014, 675-677, 880-885.
[6] Feng L.H., Zhang X.C., Luo G.Y., Application of system dynamics in analyzing the carrying capacity of water resources in Yiwu City, China[J]. J.macis, 2008, 79(3), 269-278.
[7] Lou Y.H., Kang S.Z., Cui N.B., Yang H.X., Establishment and application of comprehensive evaluation model for water-saving development level of irrigation management in Sichuan province[J]. J.totcsoae, 2014, 30(4), 79-89.

Shi Gao, Guomin Zuom, Junxiang Chen, Liming Zhou*,
Jianmei Zhou, Jiwei Liu

Study on the TD-GC-MS to Detect HD in Exhaust Gas

Abstract: The thermal desorption- gas chromatography-mass spectrometry (TD-GC-MS) determination of mustard (HD) in exhaust gas methods were studied. The experimental conditions of TD-GC-MS detection in exhaust gas of HD was optimized, making the working curve of TD-GC-MS detection in exhaust gas of HD. The effects of hydrocarbon and polyaromatic hydrocarbons (PAHs) which usually produce destruction of the abandoned chemical weapons by Japan in China (JACW) in exhaust gas on determination of HD were Investigated. The limit of detection of HD and the recovery rate of different HD concentrations on the detection of exhaust gases and under different conditions were studied. The results show that the method of HD in ambient air of a detection limit is 1.3×10^{-6} mg/m^3, this method under the condition of the detection limit of HD is 5.7×10^{-4} mg/m^3 (sampling volume 30 L).

Keywords: thermal desorption- gas chromatography-mass spectrometry (TD-GC-MS), mustard, exhaust gas, destruction of the abandoned chemical weapons by Japan in China.

1 Introduction

Blister agent is a chemical which can cause serious blister canker following exposure to the skin, in addition it can cause systemic poisoning phenomena by respiratory tract absorption. Although this type of agent causes a low mortality rate during war, the enemy has a severe loss of the battle, as it greatly increases the burden of the enemy's medical support. Mustard gas (HD) is a typical representative of this kind of agent, an alkylating agent that can cause short-term and long-term toxicity, one of the main components of abandoned chemical weapons by Japan in China (referred to as "JACW"), HD droplets to exposed human half lethal dose (LD_{50}) was about 100 mg/kg, steam/aerosol contact exposed half lethal concentration (LCT_{50}) for 15000 mg min/m^3 [1].

At present, the set method applied for mobile destruction within the territory of China, is the destruction of all types of yellow agent and therefore we must monitor HD content in exhaust gases. JACWs atmospheric emission limit concentration of HD

***Corresponding author: Liming Zhou,** Institute of Chemical Defense, PLA, Beijing, China,
E-mail: zjzlm06@sina.com
Shi Gao, Guomin Zuom, Junxiang Chen, Jianmei Zhou, Jiwei Liu, Institute of Chemical Defense, PLA, Beijing, China

is 0.002 mg/m^3. Detection of HD standard methods using organic solvent absorption, gas chromatography- pulsed flame photometric detector (GC-PFPD) [2,3], When sampling 30 L, the minimum detection limit was 0.0002 mg/m^3. Although an order of magnitude lower than the limit standard, but there are sample preparation, long analysis cycle, sample components easy to drawbacks and so on. We needed to develop more sensitive and more effective method to detect HD in exhaust gases. TD-GC-MS is a kind of method analysis of volatile organic compounds in air or soil effective detection [4-6]. A significant advantage of the thermal desorption is in front of the analysis of preconcentration analyte without dilution. The purpose of this article is to use TD-GC-MS building an effective analysis method of HD in exhaust gases.

2 Experimental Part

2.1 Main Instruments and Reagents

Thermal Desorption Instrument, MARKES company; GC-MS, Agilent 7890A-5975; Intelligent Air Sampler, Qingdao Laoshan Institute of Applied Technology; acetonitrile, chromatographically pure; 0# diesel oil, Gas station; Polynuclear Aromatic Hydrocarbons Mix, ANPEL Scientific Instrument(Shanghai)Co.,Ltd.; HD>98%, Provided by schedule 1 chemical synthetic laboratory 10 kg of ICD of PLA.

2.2 Experimental Methods and Conditions

1. *Experimental method:* First, use the portable air sampler connected aging of Tenax adsorption gas samples directly, flow 500 mL/min, sample time 60 min; then inject 1 ul acetonitrile target substance into aging through the Tenax adsorption tube with a flow rate of about 500 mL/min of N2 purge for 2 min in the laboratory; then take the Tenax sorbent tube into TD (thermal desorption), open test procedure, complete the test.
2. *Condition of TD-GC-MS:* TD detect the target process generally divided into the following three stages: pre-desorption, Tenax tube-desorption, trap-settings. GC-MS condition: injector temperature 280°C, chromatographic column (HP-5MS, 30×0.250 mm, 0.25 Micron); temperature programming: The initial temperature of 60°C, temperature rate 20°C/min, up to 280°C, keep 3 min, column flow rate of 1 mL/min, assist the heater temperature: 280°C

3 Results and Discussion

3.1 Optimization the Condition of TD

1. *Optimization of the desorption temperature of Tenax tube:* TD-GC-MS detect the target process generally divided into the following three stages: Tenax tube-desorption,trap-settings. The purpose of pre-desorption is primarily purging at room temperature for a period of time to exclude interference of volatile organic compounds and water vapor in Tenax tube; desorption process mainly includes the desorption time and desorption temperature two parameters, desorption process of the heating rate is solidified in the program and is relatively slow, so the warming of the desorption process needs certain time (about 2.5 min); TD process is under the maximum heating rate (100°C/s) up to the set temperature and maintains a certain period of time, samples at the same time through the instantaneous temperature into steam into the subsequent monitoring in pipeline and then into the GC–MS. In this section, we investigate the effects of desorption temperature, that is, other conditions remain unchanged, only change the desorption temperature. The specific process is pre-desorption for 2 min, desorption 3 min, desorption temperature set 200°C, 240°C, 280°C, 320°C respectively; The initial temperature of cold trap is -10°C, The maximum heating rate to 280°C, holding 1 min. The results are shown in Fig. 1.

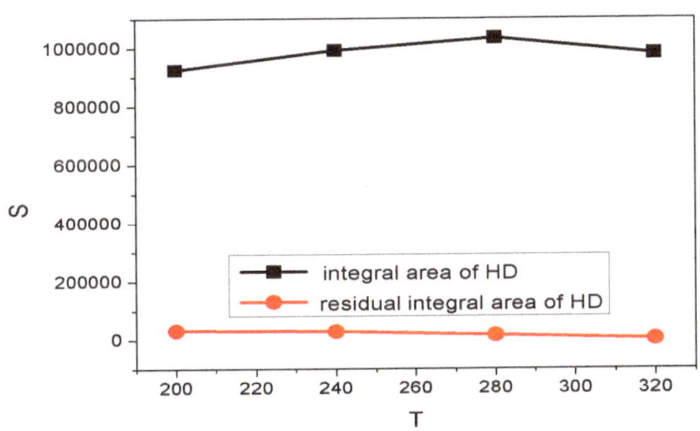

Figure 1: The effect of Tenax tube desorption temperature change

As shown in Fig. 1, with the increase of desorption temperature, the HD peak area appears to increase gradually, and residues of Tenax tube have gradually decreased. From the GC-MS peak also can be seen that with the increase of desorption temperature, HD's peak tends to be more and more sharp. The reason may be that with the increase of temperature, HD desorption efficiency increased in the Tenax tube. But when the temperature is too high, HD peak area showed smaller trend again, reason may be that the temperature is too high, decomposition occurs in the HD.

Through the above experiment, when the desorption time is 3 min, sets the desorption temperature to 280°C will be good for desorption of HD.

2. *Tenax tube desorption time optimization:* In this section we look at the influence of Tenax tube desorption time, other conditions maintain constant, only the desorption time changes. The specific process is pre-desorption 2 min; the temperature of desorption is 280°C, desorption time set 1 min, 3 min, 5 min, 8 min respectively; the initial temperature of cold trap is -10°C, the maximum heating rate to 280°C, holding 1 min. The results are shown in Fig. 2.

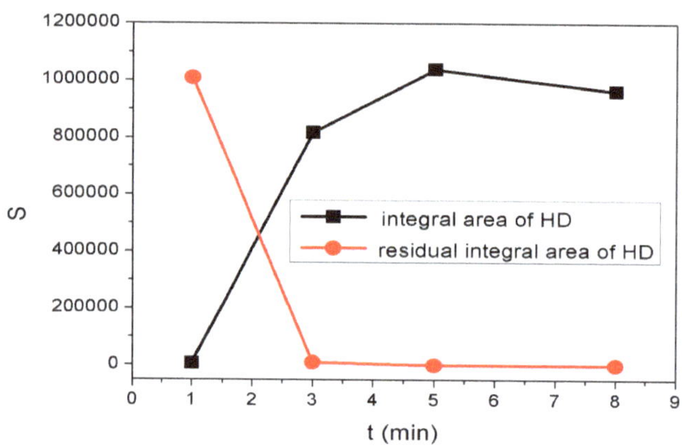

Figure 2: The effect of Tenax tube desorption time change

As shown in Fig. 2, we can clearly identify that with the extension of Tenax tube desorption time, desorption efficiency of HD is obviously rising, HD in the Tenax tube residue showed a gradual dereasing trend. When desorption 1 min, HD has lowest desorption efficiency. The main reason is that the Tenax tube heating rate is limited, up to 280°C takes about 2.5 min, but the desorption time only set 1 min, so when the adsorption process is completed, the temperature is not to set temperature, lead to most of HD residues in Tenax tube. When the desorption time was 8 min, the desorption efficiency of HD has lowered, perhaps the main reason is that the high temperature is kept for too long, causing a certain degree of decomposition of HD.

In conclusion, Tenax tube after adsorption HD desorption condition is: desorption time 5 min, desorption temperature 280°C.

3. *Cold trap thermal stripping temperature optimization:* The cold trap heat stripping time 1 min unchanged, set the highest temperature of cold trap to 200°C, 240°C,

280°C and 320°C, heating rate is the largest, inspects the change of cold trap temperature heat stripping effect on HD thermal desorption.

As shown in Fig. 3, when HD through Tenax tube after desorption to cold trap, keeping the cold trap heat keep to 1 min, cold trap instantaneous temperature from -10°C up to 200°C, 240°C, 280°C and 320°C, we can see from figure 3, the change of temperature of cold trap has no impact on HD desorption. Based on the physical characteristics of HD, the temperature of cold trap can be set at 280°C. In addition each temperature has a certain degree of residual, the possible reasons are as follows: one is the time to maintain heat of cold trap relatively short; second, the cold trap instantaneous temperature desorption HD's ability is limited, the present basic has reached its limits, the dose will have certain residue.

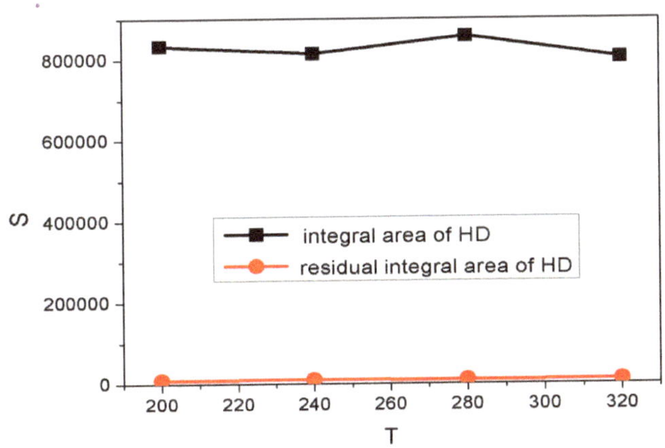

Figure 3: The influence of changing temperature of TD cold trap

4. *Optimization the maintain time of Cold trap thermal desorption:* Other conditions remain unchanged, cold trap thermal desorption keep time set respectively for 1 min, 3 min, 5 min, investigate the influence of high temperature holding time of cold trap on HD desorption. As shown in Fig. 4.

From the above, we know to keep the temperature of cold trap constant at 280°C, maintain the time of the high temperature set to 1 min, 3 min, 5 min. At 3 min, HD desorption efficiency is highest, at the same time, every time there is a residual. This fully shows that the cold trap instantaneous desorption, and limited ability to HD when HD amount reaches a certain value, change the temperature of cold trap or change the time of cold trap maintain high temperature, residual phenomenon is inevitable.

To sum up, cold trap thermal stripping HD can be set to the condition of temperature 280°C, holding 3 min. At the same time we should pay attention to the intake of HD dose, avoid the generation of residual phenomenon.

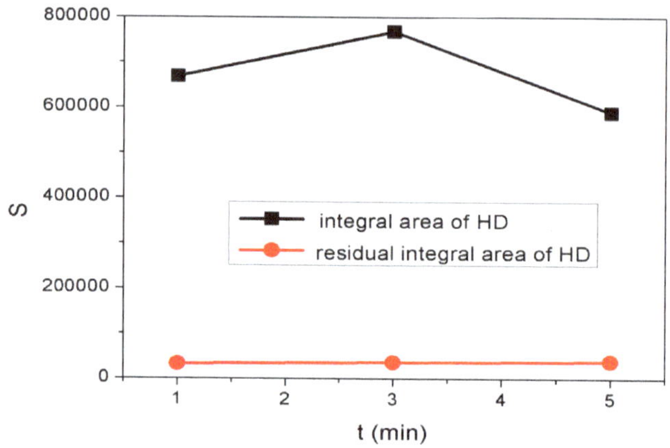

Figure 4: The influence of high temperature holding time of TD cold trap on HD desorption

3.2 Standard Curve of TD-GC-MS Detection of HD

Taking 1 μl concentration of 0.1134 mg/mL, 0.2268 mg/mL, 1.134 mg/mL, 11.34 mg/mL, 113.4 mg/mL, 1134 mg/mL HD acetonitrile solution into Tenax tube, the TD and GC-MS in various conditions unchanged, fitting the HD quality inject to Tenax tube and GC-MS monitoring to the peak area of HD and get a TD - GC - MS detection HD, as shown in the Fig. 5 below.

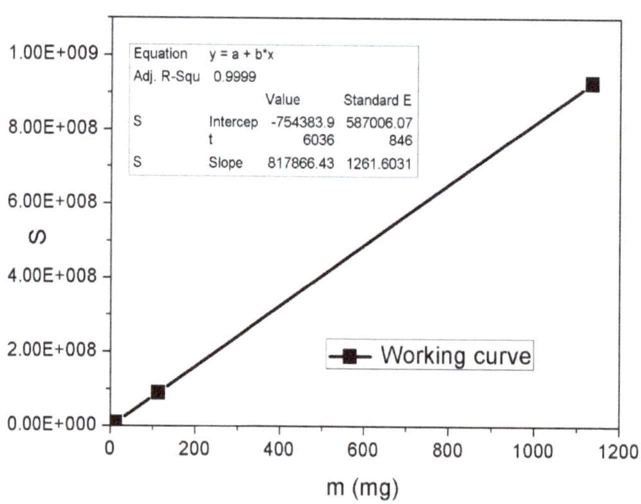

Figure 5: Work curve of TD-GC-MS detect HD

As shown in Fig. 5, TD-GC-MS detect HD work curve linear relationship is very good, fully meet the requirements, through linear fitting R^2 value 1. This fully proves the TD - GC-MS detection accuracy of HD in the air.

3.3 The Effects of All Kinds of Compounds in the Air in TD-GC-MS to Test HD

Due to the large number of existence and hydrocarbons in the air, but the destruction of the site contains a large number of secondary pollutants benzene class material, so this article will be 0 # diesel oil as the representative of the hydrocarbons in the air will contain a mixture of 16 PAHs in the standard sample as a representative of the containing benzene ring type matter. Investigation effects of TD-GC-MS test HD.

3.4 The Effects of HYDROCARBON in TD-GC-MS to Test HD

This section investigates the effects of the diesel and HD concentration ratio of 10:1, 100:1, 1000:1 three cases to TD - GC - MS detect HD respectively. Experiments found that when the diesel and HD concentration ratio of 1000:1, TD - GC - MS has been unable to distinguish between the existence of HD, diesel with HD concentration ratio of 100:1, TD - GC - MS can clearly distinguish the HD. As shown in the figure below.

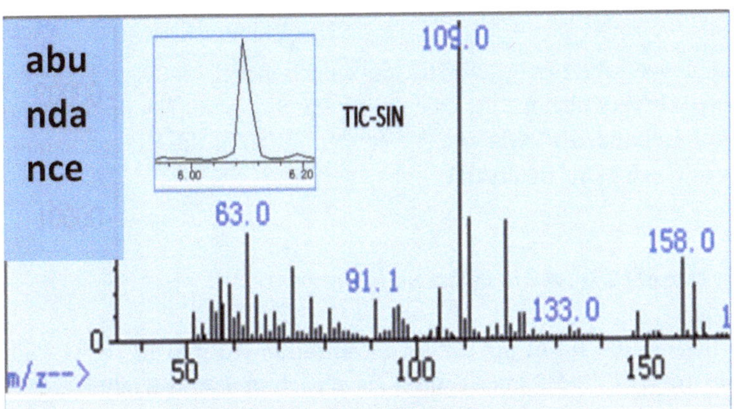

Figure 6: The effects of Hydrocarbon in TD-GC-MS to test HD

It can be seen from the above, diesel concentration with HD concentration ratio of 100:1, TD - GC - MS detection for HD is very obvious. By calculating the signal-to-noise ratio(S/N) is 180. The implied, the diesel oil concentration and the concentration of the HD than about 500:1 is limit of TD - GC - MS detect HD.

3.5 The Effects of PAHs in TD-GC-MS to Test HD

This section take HD and contains 16 kinds of PAHs and injection of mixed into Tenax tube, investigates the effects of PAHs and HD concentration ratio of 10:1, 200:1, 2000:1 three cases to TD - GC - MS detect HD respectively. Experiments found that when the diesel and HD concentration ratio of 2000:1, TD - GC - MS has been unable to distinguish between the existence of HD, diesel with HD concentration ratio of 200:1, TD - GC - MS can clearly distinguish the HD. As shown in the figure below.

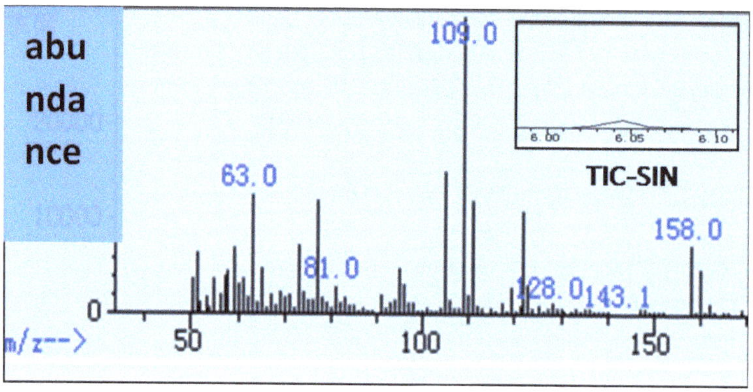

Figure 7: The effects of Hydrocarbon in TD-GC-MS to test HD

Can be seen from above, when PAHs and HD concentration ratio of 200:1, TD - GC - MS detection for HD is very obvious. By calculating the S/N is 90. The implied, the concentration of diesel and HD PAHs and the concentration of the HD than about 500:1 is the limit of TD - GC - MS detect HD.

3.6 The recovery rate of TD-GC-MS to detect HD in the air

With trace injection needle 1ul of the same concentration HD into 10 Tenax tube respectively. Keep the TD - GC - MS same conditions of each step. Investigation of TD - GC - MS detect the recovery rate of HD in the air. As shown in the figure below.

The chart shows that under the condition of low concentration, due to sampling error, instrument error reasons, such as HD recovery rate volatility relative to a few bigger, show the relative standard deviation is larger, about 5.6%; at high concentration, HD recovery is less volatile, relative standard deviation value is small, about 1.6%. This fully verified by Tenax tube adsorption desorption HD to detect HD method is feasible in the air.

Table 1: The recovery rate of TD-GC-MS to detect HD

N	AQ (ng)	RQ (ng)	RR %	RSD %	AQ (ng)	RQ (ng)	RR %	RSD %	AQ (ng)	RQ (ng)	RR %	RSD %
1	0.1134	0.1104	97.39		1.134	1.129	99.57		11.34	10.99	96.87	
2	0.1134	0.1238	109.18		1.134	1.183	104.30		11.34	11.50	101.42	
3	0.1134	0.1136	100.16		1.134	1.227	108.22		11.34	11.38	100.35	
4	0.1134	0.1154	101.79		1.134	1.137	100.30		11.34	11.58	102.08	
5	0.1134	0.1234	108.82		1.134	1.102	97.17		11.34	11.10	97.84	
6	0.1134	0.1264	111.48	5.66	1.134	1.037	91.42	5.49	11.34	11.21	98.84	1.63
7	0.1134	0.1150	101.42		1.134	1.068	94.16		11.34	11.41	100.63	
8	0.1134	0.1142	100.69		1.134	1.078	95.06		11.34	11.26	99.25	
9	0.1134	0.1181	104.18		1.134	1.216	107.20		11.34	11.30	99.69	
10	0.1134	0.1315	116.00		1.134	1.177	103.80		11.34	11.18	98.59	

(N: Number; AQ: Adding quantity; RQ: Recycled quantity; RR: Recovery rate; RSD: Relative standard deviation)

3.7 The Detection Limit of TD-GC-MS Detect HD in the Air

1. *The detection limit of clean Tenax tube detect HD:* Experiment by using different doses of HD solution into clean Tenax tube, the TD and GC - MS condition unchanged, inspects the HD detection limit. Below for HD injection of 113.4 pg, TD - GC - MS of TIC and MS figure, through the signal-to-noise ratio (S/N) detection, the S/N value is 90. In general, the default for the quantitative detection of S/N limit of 30, through calculation, canister to clean Tenax detection in HD the detection limit of 1.3×10^{-6} mg/m^3.
2. *The detection limit of working condition of Tenax tube detect HD:* Experiment by using different doses of HD solution has been injected into the day to day work environment sampling Tenax tube, the TD and GC - MS condition unchanged, inspects the HD detection limit. Below for HD injection of 11.34 ng, TD - GC - MS of TIC and MS figure, through the signal-to-noise ratio detection, the signal-to-noise ratio of the S/N value is 200. In general, the default for the quantitative detection of signal to noise ratio limit of 30, through calculation, working condition of Tenax detection in the canister about detection limit was 5.7×10^{-4} mg/m^3 of HD.

4 Conclusions

Based on exhaust gas in HD TD - GC - MS detection method research, optimizes the TD method to detect gas in HD - GC - MS experimental conditions, making the TD - GC - MS detection in the exhaust gas HD method of the work curve, and examines the hydrocarbons in the air, and the repeated field often exist PAHs in the exhaust test the effect of HD in the exhaust gas of this method. When hydrocarbons and PAHs in the exhaust gas concentration with HD concentration ratio of 500:1, the method is used to inspect the exhaust HD has no effect, finally the method detecting waste gas was investigated, the recovery rate of different concentrations of HD. The results show that the method of HD in ambient air of a detection limit is 1.3×10^{-6} mg/m^3, this method under the condition of the detection limit of HD is 5.7×10^{-4} mg/m^3 (sampling volume 30 L).

References

[1] McGuire, R., Kirvel, R.D., Decontamination issues for chemical and biological warfare agents: how clean is clean enough? Int. J. Environ. Health Res. 2001, 11: 128.

[2] The atmospheric pollutants emission standard of destruction of Japanese abandoned chemical weapons in China,GB19060-2003, Beijing.

[3] Destruction of Japanese abandoned chemical weapons in China Exhaust gas chromatography for determination of the mustard gas in (trial),HJ/T 105-2003, Beijing.

[4] Rongrong PAN,Ganglian QU,Guohua MA, Central composite design method applied to the determination of the thermal desorption VX [J]. Journal of chemical defense, 2008, 4:19-21.

[5] Muir B., Quick S., Analysis of Chemical Warfare Agents II Use of Thiols and Statistical Experimental Design for the Trace Level Determination of Vesicant Compounds in Air Samples[J]. J. Chromamgr. A,2005. 1068:315-326.

[6] Zhou J.M.,Zhou L.M.,Chen J.X.,Thermal desorption technology in the destruction of chemical weapons environmental monitoring application prospect, CEPPH 2012, 155-157.

Shixian WANG and Xia LU*

Study on the Regeneration of Waste Lubricating Oil by Porous Nano Adsorption

Abstract: Recycling of waste lubricating oil, not only can prevent oil pollution, environmental protection, but also can save limited oil resources for human beings. On the basis of domestic and foreign analysis and research on waste lubricating oil regeneration and adsorption process, the preparation method of porous Nano additives and the regeneration process of waste lubricating oil are put forward. The results showed that the size distribution of the nanoparticles was 10~15 nm, and the particles were loose aggregation, with porous structure, the specific surface area of the product was 63.74 m^2g^{-1}, the pore size distribution of the product was disordered, and the adsorption performance was good.

Keywords: waste lubricating oil, adsorption process, Nano-Porous.

1 Introduction

With the development of the world economy, the application of lubricating oil has become more and more extensive; the world's average annual consumption of lubricating oil is about 4000 million Ton. China, as the world's second largest oil consumer, had an amount of lubricating oil consumption in 2015 as high as about 6000000 Ton. The recovery and regeneration of waste lubricating oil has become an urgent problem to be solved in the situation that the world energy is becoming more and more serious. In Europe, there are about 500 million Ton waste lubricants every year, of which 40%-50% release to the environment without process. The recovery of oil in our country is less than 20% of the total amount. Waste lubricating oil, few but wide diffusion, Ccause pollution on the environment e.g. atmosphere, soil and water every hour, and every minute. The shortage of energy and pollution to the environment are getting more and more attention from every country [1]. Lubricating oil is made up of 80%-90% base oil and 10%-20% additives, and the main chemical composition is a variety of hydrocarbons and a small amount of non-hydrocarbon mixture. However the performance of lubricating oil, after a period of use, deteriorate dramatically due to physical, chemical, or human factors, and some impurity contamination generate such as aldehydes, ketones, resin, asphaltene colloidal materials, carbon black and

***Corresponding author: Xia LU,** The College of Post and Telecommunication of Wuhan Institute of Technology, Wuhan City, Hubei Province, PR China, E-mail: 15926373810@163.com
Shixian WANG, The College of Post and Telecommunication of Wuhan Institute of Technology, Wuhan City, Hubei Province, PR China

organic acid, salt, water, metal scraps [2]. Therefore, lubricating oil can no longer continue to be used and become waste lubricants. In fact, the real deterioration is only a few percent of them [3]. Therefore, how to effectively remove these impurities in the waste oil, is the key to regeneration of waste oil [4].

At present, the majority of the waste lubricating oil regeneration process in our country is still following the traditional sulfuric acid clay regeneration process. The use of concentrated sulfuric acid will generate a considerable amount of three wastes e.g. acid residue, acid gas and water. Regenerating Lubricating oil used in mechanics equipment produce 150-250 kg per ton of acid slag, waste diesel oil produce more acid residues, up to 400-600 kg per ton. The regeneration of the traditional process will be accompanying new waste and environmental pollution. Under the condition of saving oil resources and increasing demands for environmental protection, high efficiency, low cost, less pollution or no pollution of the waste oil recycling process has received more and more attention. In order to avoid the waste oil regeneration cause second pollution to the environment, the main approach to improve the regeneration process is to improve the operating conditions and equipment, to reduce acid residue production, and to achieve comprehensive utilization. The low acid refining process in foreign countries should be extracted with propane. The refining process needs a large amount of treatment equipment, high investment, harsh operating conditions, and requires skilled technicians and workers to operate it. It is obviously not suitable for China's national conditions, and difficult to copy. At present, all research of acid free refining process has its shortcomings; it is difficult for comprehensive promotion in China.

2 Present Status of Adsorption Process

The deterioration process of lubricating oil is a complex physical and chemical process, which contains various types of reactions, which can be classified in two categories. The first category reaction is the type of oxidation of alkanes and cycloalkanes and aromatic hydrocarbons with long side chains: hydrocarbon → peroxide → hydroxyl acids → asphaltene acids carbides. The second category reaction is the type of oxidation of aromatic hydrocarbons with short side chain and side chain: hydrocarbon → phenols → peroxide → glial → asphalting → semi coke oil. The main products of the reaction are the condensation product of hydroxyl acid, asphalt resin, organic acid, salt, carbon and graphite, etc. The molecular structure of the depredated hydroxyl acid molecule is changed into polar molecules, and the interfacial tension decreases, which is characteristic of the oxidation deterioration of the lubricating oil.

Adsorption is a kind of process used in the purification of oil. It can remove the impurity material such as acid oxide, water and so on [5]. The micro structure of adsorbent is porous, with large specific surface area. Surface area of one gram clay can reach 100-300 square meters, and surface area of a gram silica gel get up to

300-450 square meters. In the adsorption process, adsorption agent and oil contact with each other, and the liquid asphalt, naphthenic acid, sulfonic acid, gum and a variety of unsaturated hydrocarbons and other substances are removed. Adsorption is an effective method for the decolonization of oil. Clay is the most important adsorbent in oil purification adsorption separation method [6].

Clay is a kind of bentonite, and its chemical composition includes silica, alumina, iron oxide and a small amount of calcium oxide, magnesium oxide, and other [7]. Its molecular formula is Al (OH) $_3$·nH$_{20}$·mSiO$_2$. Clay can effectively remove polar molecules such as Glioma, naphthenic acid, sulfonic acid, sulfonate. Xu Gaoyang et al. [8] studied regeneration process of the marine waste lubricating oil. After absorption by clay, the quality of waste oil improved remarkably. The regeneration condition depicts as following: the temperature is 80°C, and the stirring time is 2 h, the amount of clay is 8%-20%. At the same time, filtering of regeneration waste oil was also studied. Using the composite filter agent, the filtration rate can be improved greatly, the quality of regeneration waste oil reached the level of practical. Song Rui Zhou et al. [9], experimentally studied the regeneration of waste lubricating oil by using the short-range distillation and clay hydro finishing process. The quality of the regenerated lubricating oil has reached the new lubricating oil technical index. This studied showed that this process has a high recovery rate, short regeneration cycle, clean and environmental protection, no second pollution, and has very good economic and social benefits. Sheng Ling Zhang et al. [10], using two kinds of surfactants as flocculants and activated clay as adsorbent, treated the waste lubrication oil by flocculation and adsorption regeneration process. They determined the optimum flocculation conditions as following: the optimum temperature is 85°C; 150 waste lubricating oil, the amount of two kinds of flocculants dosage are 3.7 g and 1.3 g. Optimal adsorption condition is atlapulgite 12.3%, temperature 110°C, time 90 min. After the elemental analysis, the content of Zn, Pb, Ca, P and other elements of the refined oil was significantly decreased. Refined oil physical and chemical indicators are in line with national standards. Qingzhi Fei et al. [11] removed dissolved water emulsion by AG emulsification water absorbent. After the clay adsorption treatment, the waste lubricating oil was regenerated. The best operating conditions of waste lubricating oil in two factories were studied. After adding 1% AG agent and 6% clay into the presedimentation waste oil, mixing 20 min at 70°C, and static deposition 24 h, the regenerated oil, meeting the physical and chemical indicators, can be obtained. The process has the advantage of no pollution, high yield, low operation cost, and have been transferred to production. Shuiqing He et al. [12], used fly ash as a substitute for other adsorbents, and added different amounts of fly ash into waste lubricating oil with different labels. Waste lubricating oil was regenerated and the improving effect is very significant. The process can not only reduce the production cost and labor intensity, but also improve the important indexes of the regenerated oil. Fly ash for a large number of waste lubricating oil recycling can achieve waste treatment. The effect is significant, and the process has a certain promotion prospects. Flow chart of the process described as following: waste oil → coarse filtration → fine filtration → heating → stirring → Adding fly ash → settling → dewatering → sewage → fine filtration → finished products.

In the process of regeneration of waste oil, frequently used adsorbent include clay, silica, activated carbon, bauxite, alumina and so on. The flow chart of purification and regeneration of waste lubricating oil mainly based on adsorption is shown below: waste oil → natural precipitation → adsorption → filtration → reconcile → product oil. Compared with the traditional process, the process of adsorption and regeneration of waste lubricating oil has the following advantages.

1. The efficiency of the regenerated lubricating oil is high, the technological process is simple and the regeneration period is short;
2. The waste lubricating oil has no acid residue, no waste water and no waste gas, and can reduce the pollution, and can achieve a good environmental protection effect.
3. The residue produced after regeneration can be used to reconcile the pitch black, so that the resources can be fully utilized.

Because the comprehensive effect of the ideal adsorbent is very seldom, the purpose of this paper is to configure a new adsorbent -porous micro Nano additive to regenerate lubricating oil. The physical and chemical properties index was achieved after adding porous micro Nano additive, dehydration, polymerization of slag, precipitation and cleaning.

3 Porous Nano Additive

Porous material is a new material system developed in twentieth Century. It includes metallic porous materials (i.e., foam metal) and nonmetallic porous materials (i.e., foam plastics and porous glass, etc.). The structure is characterized by regular arrangement, adjustable size of the pore structure, high specific surface area and large adsorption capacity, and the porous materials are widely used in the fields of macromolecular catalysis, adsorption and separation, Nano material assembly and biochemistry. In recent years, the research of porous materials has been carried out in all directions, so far, the research in this field has made remarkable progress.

In view of the situation of China's oil source dispersion, serious hybrid, multi species, small quantity, unstable quality of raw materials, a new kind of porous Nano additive was proposed. The new Porous micro Nano material additives can be directly added to waste lubricating oil, and most of waste oil's physical and chemical properties can be recovered after dehydrated, aggregate out of slag, precipitation, cleaning. This recycling process doesn't need waste oil collection, refining, and can not cause any pollution to the environment. Porous Nano additives are obtained by water and heat and solvent thermal method, the porous Nano additives have advantage of larger specific surface area, strong ability to absorb water and super adhesion. Wide application of porous micro Nano additives will effectively prolong the service life of lubricating oil and oil cycle. Without waste oil recycling, no need of the pickling refining, dispense with the refining process and regeneration costs substantially reduced, avoiding the secondary pollution, these characters are the true sense of environmental protection and energy saving projects.

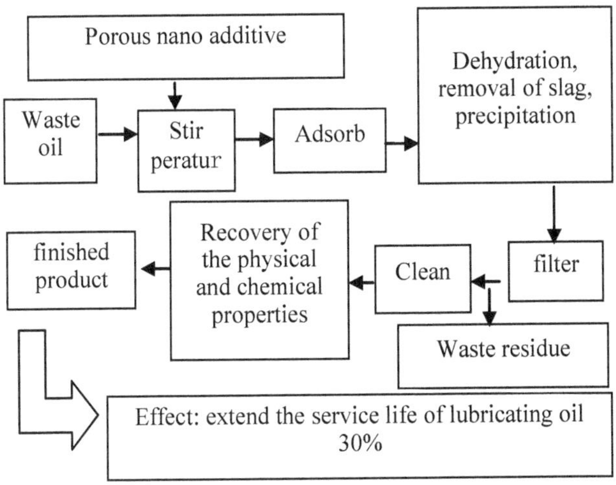

Figure 1: Porous Nano-additives process chat

4 Experiment

Test material:
Attapulgite ($Mg_5Si_8O_2O(OH)_2(OH_2)_4 \bullet 4H_2O$); Surface modifying agent Two methyl phosphoric acid four ethyl ester, three amine, cyclohexane, distilled water, anhydrous alcohol;

Test equipment:
Stainless steel reaction kettle lined with PTFE, digital oven, drying box

Procedures and methods for the preparation of porous Nano additives are as follows:

1. The choice of reaction precursor, determine the reaction precursor ratio;
2. To explore the precursor to join the order, mixing;
3. Packed Kettle, Sealed kettle, put into the oven;
4. To determine the reaction temperature, time, state (static or dynamic crystallization) reaction;
5. Taken Kettle, cooling (air cooling or water cooling), sampling;
   Filtering, washing and drying.

5 Results and Discussion

5.1 Microscopic Characterization of Nanoparticles

Figure 2 shows the TEM photo and its diffraction pattern of the prepared copper nanoparticles.

It can be valuable that Nano particles are spherical particles, the size of the nanoparticles is distributed in 10~15 nm, and the particles are loose.

Figure 2: TEM image and diffraction spectrum of the copper nanoparticles

5.2 Adsorption Properties of Nano Copper

Figure 3 is the N2 isotherm adsorption curve of porous Nano particles. The specific surface area of the product was calculated according to the BET model. The area is 63.74 m^2g^{-1}, which shows that it has a porous structure. The pore size distribution of the product is also disordered, which shows that the Nano particles are disordered mesoporous materials, and the adsorption performance is better.

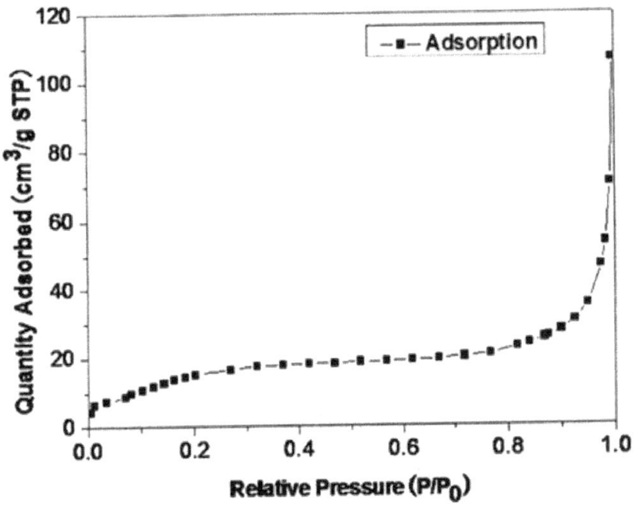

Figure 3: Isothermal adsorption curve of porous Nano particles N2

6 Conclusion and Prospect

1. Porous Nano additive adsorption regeneration process is simple, easy to use. The regeneration cycle is short, meanwhile overcome the oil source dispersion, recycling difficult, and a series of problems.
2. The waste lubricating oil regenerating process, without the process of pickling, alkali neutralization, water washing and the like, will not produce acid residue, waste water and waste gas, and can reduce the second pollution, and can achieve a good environmental protection effect.
3. Domestic and foreign technology is moving towards the direction of no pollution, environmental protection. The development of waste oil suitable for China's environmental protection and economic regeneration process is an urgent research topic. Adsorption process will be one of the more important research directions.

Acknowledgment: This work is financially supported by the Science and technology guiding project of Hubei Province (No.B2016486).

References

[1] Yang H.W., Fei Y.W., Hu J.Q., Regeneration of waste lubricating oil at home and abroad [J], lubricating oil., 2006, 21 (6), 9-11.
[2] Mendelevich A., Zamotailov V. V., Sarganova R S., et al., Use of centrifuges and separation of turbine oils[J], Chemical and Petroleum Engineering., 2002, 38, 5-6.
[3] Hey Zhang, Review on the present situation of recycling and utilization of waste lubricating oil in mine [J], Chinese materials recycling, waste mineral oil., 1996, (6), 22-24.
[4] Zhang X.M., Jiao Z.J., Jia Y.M., et al., Regeneration technology of waste lubricating oil for industrial pollution [J], environmental science and technology., 2008, 31 (3), 39-42.
[5] Khalajabadi, S.Z., M.R. Abdul Kadir., S. Izman., et al., Synthesis, microstructure and biodegradation behavior of nano-Si and nano-ZnO/Si coatings on a Mg/HA/TiO_2/MgO nanocomposite[J], Ceramics International., 2015,41(9, Part A), 11346-11358.
[6] Department of information science and technology., Institute of petroleum and Chemical Research Institute of petroleum and chemical industries[M]. Beijing: Petroleum Industry Press, 1978:21-80.
[7] Shao D., Tang D., Yang J., et al., Nano-structured composite of Si/(S-doped-carbon nanowire network) as anode material for lithium-ion batteries [J],Journal of Power Sources., 2015, 297, 344-350.
[8] Xu G.Y., Lu M.., Regeneration of marine waste lubricating oil [J], China Resources Comprehensive utilization., 2003, (12), 14-16.
[9] Song R.Z., Yin Y.S., Wang Y.Y., et al., Short path distillation technology in the regeneration process for waste lubricating oil application [J], Chemical progress., 2006, 25 (L1), 1371-1374.
[10] Zhang S.L., Liu H.W., Zhao X.G., Study on the process of flocculation and adsorption regeneration of waste lubricating oil [J],chemical world., 2002, (10), 527-529.
[11] Fei Q.Z., Liu W.D., Xu Z., et al., Experimental study on new technology of regeneration of waste lubricating oil [J], Liaoning chemical industry., 1996, 146 (6), 26-27.
[12] He S.Q., Application of fly ash in waste lubricant regeneration [J],Comprehensive utilization of resources in China., 2002(7), 23-24.

B. Chen, L.X. HAN*, L.l. Yuan

The Determination and Application of Characteristic Parameter for Spilled Oil Vertical Concentration Distribution Formula

Abstract: After analyzing and summarizing previous research, it is obvious that the characteristic parameter, surface spilled oil concentration, is the key limiting condition for the usage of the spilled oil vertical concentration distribution formula, because there are no appropriate field detection condition and measurements. However, the vertical concentration distribution should receive more attention in the environmental risk assessment in consideration of the location of the aquatic organism and the water intakes. Based on the formula of vertical concentration distribution, a mathematic calculation method of the characteristic parameter, spilled oil surface concentration, was put forward, which mainly applies to the stage of surface tension and viscous force for marine oil spill. That is to say we establish the response relationship of the vertical concentration and the surface concentration under the most unfavorable situation by integrating along the vertical direction to get the average concentration. Finally, the key characteristic parameter, surface concentration, is available and it can be used to predict the characteristics of the vertical concentration distribution, which will provide water pollution characteristic parameters for the emergency warning of sea water quality.

Keywords: oil spill, vertical concentrate distribution, surface concentration, emergency warning.

1 Introduction

Recently, accidental oil spills are more frequent and on a larger scale. Oil spills in aquatic environment have become more common because of growth in marine oil production and transport over the last few decades [1,2]. The petroleum pollutants, leak into the water body, will cause significant damage to social economy, and also bring serious harm for regional water environment and ecological environment.

Oil may be released into the marine environment as a result of oil extraction activities such as drilling, storage and transportation [3,4]. Such as, the oil tankers may cause oil spill and make significant environmental problems in the event of

***Corresponding author: L.X. HAN,** College of Environment, Hohai University, Nanjing, China, E-mail: hanlongxi999@163.com
B. Chen, College of Environment, Hohai University, Nanjing, China
L.L. Yuan, National Center of Oceanographic Standards and Metrology, Tianjin, China

collisions or grounding [5,6]. In 1989 Exxon Valdez oil spill discharged 260 thousand barrels crude oil in Prince William Sound, Alaska through the ship's ruptured hull [7]. Some scholars found that the oil drowned into the water column could influence the water environment persistently.

Once entering into the water, the spilled oil spreads by the action of wind or currents and forms a surface slick covering a large area of the water surface [8,9]. Some light hydrocarbons and some polar components begin to go into solution in the underlying water column, even though most of these are lost through evaporation into the atmosphere [10,11]. Moreover, some of the oil is emulsified into the water column as small dispersed droplets [12-14]. These droplets may become dispersed because of the action of currents, or they may become attached to suspended particulate matter and slowly settle to the bottom. The turbulence action can also cause water to become entrained in the oil forming water-in-oil emulsion,which may eventually form dense tar balls. It is clear from these time scales that the oil undergoes significant modification during the first few hours following a spill.

Many of the oil spill models developed during the last decade only simulate the advection and spreading processes. However, what draws more attention are the trace of migration, surface distribution of oil spill and thickness of oil film [15-18]. The concentration of oil spill, especially vertical concentration, is not enough.

Several studies were performed to research the vertical oil dispersion in a wave field through experimental tests. Mackay et al. [19,20] conducted an experimental model and proposed equations to calculate the amount of spilled oil intruded into the water column due to natural oil dispersion. Otherwise, the surface oil dispersion was investigated by Delevingne and Sweeney in a small laboratory flume as well as in a test basin [21]. Their study resulted in an empirical equation to estimate the oil dispersion rate as a function of oil type and breaking wave energy. Also, an equation was proposed to predict oil droplet size distribution. In addition, an experimental program is conducted by Reza Parsa et al. [22] to investigate the vertical oil dispersion of surface oil spills in a regular wave field, and which determines the trends of oil concentration variations due to natural oil dispersion and formulate the magnitude and time of maximum oil concentration.

Based on the mechanism of oil droplets formation and diffusion, Zhao [23] established a differential equation for oil droplets turbulent diffusion and put forward spilled oil vertical concentration distribution formula. But how to determine the characteristic parameter, datum point concentration, is unknown. Blumer et al. [24,25] get a certain hydrocarbon compound concentration by computing the integral for the vertical diffusion equation. What is more, it is necessary to have the saturated solubility of this hydrocarbon compound in the sea. However, it is difficult to use this method for petroleum pollutants, as such pollutants contain so many kinds of hydrocarbon compounds. Up to now, rare equation is proposed to estimate oil concentration in the water column and most of derived equations evaluate the rate or amount of oil dispersed in the water. Furthermore, no appropriate field measurements or experimental data is collected for numerical models validation [26]. Actually, marine biological communities located at different depths, and almost all the water intakes are set under the water surface with a certain depth. Therefore, it is significant to determine petroleum pollutants concentration distribution in the vertical direction succinctly. And the vertical concentration distribution should also receive more attention in the environmental risk assessment.

Previous research mainly focused on the total amount of oil drowned in the water column, and therefore no information was available regarding oil concentration in the water column. In particular, most of the spilled oil had dissolved and emulsified into the water in the stage of surface tension-viscous force according to some studies [27,28]. Under this premise, according to the formula of vertical concentration distribution, a mathematical method is conducted to determine the vertical oil dispersion of surface oil spills in this study. The objective is to determine the vertical oil concentration variations due to natural oil dispersion and formulate the magnitude of oil concentration. The correlation between oil concentration and some important parameters such as thickness of oil film, amount of spilled oil into the water and the vertical distance are indicated and also oil concentration variations are quantified.

After analyzing and summarizing the previous research conclusions, it is obvious that the characteristic parameter, surface spilled oil concentration, is the key limiting condition for the usage of the spilled oil vertical concentration distribution formula. However there are no appropriate field detection conditions and measurements. Therefore, this paper puts forward a mathematical calculation method for computing the surface spilled oil concentration, which is also regarded as the datum point. That is to say we establish the response relationship of the vertical concentration and the surface concentration under the most unfavorable situation by integrating along the vertical direction to get the average concentration. After a series of calculations, the surface concentration is available. Then, the petroleum concentration of different depth could be predicted so that we can evaluate the accidental influence on the marine biological communities and water intakes which are located in different depth. The characteristic parameter calculated by this mathematical method fills in the gap between theory and practice, making the usage of the spilled oil vertical concentration distribution formula possible. It will be convenient for the emergency warning of water quality and the biomass loss calculation.

2 Concentration Vertical Distribution of Oil Spill

Through analyzing the mechanism of oil droplets formation and diffusion, Zhao et al. [23] established differential equation for oil droplets turbulent diffusion and earlier proposed an opinion that this is to reduce along the depth in the water body. The variation of oil concentration with water depth due to difference droplet size and diffusion coefficients pictured in Figure 1 and Figure 2 respectively. Then, this opinion, which the vertical concentration obeys index distribution, was further confirmed by field testing, and also expressed by Figure 3 [29]. Now, this point, the vertical concentration is exponential to reduce along the depth in the water, is widely recognized.

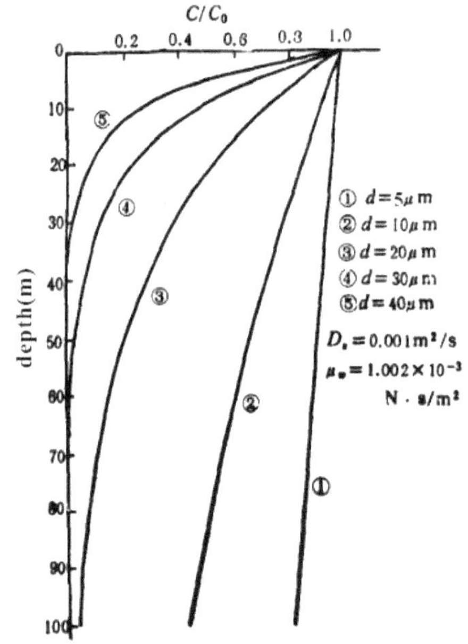

Figure 1: Variation of oil concentration with water depth due to difference droplet size by Zhao et al. 1990

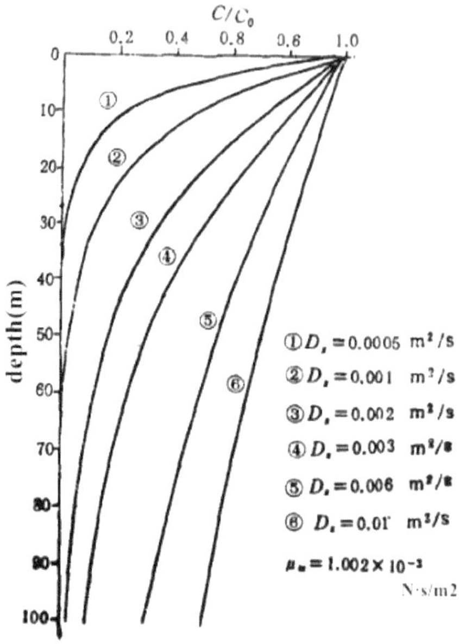

Figure 2: Variation of oil concentration with water depth due to difference diffusion coefficients by Zhao et al. 1990

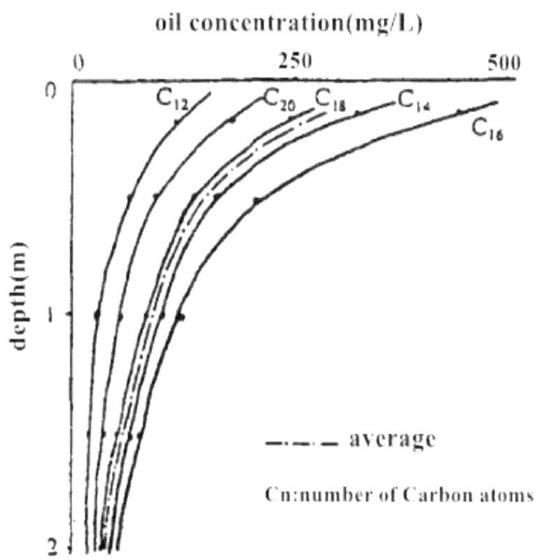

Figure 3: Vertical distribution of n-alkane in the sea by Yang et al. 1992

Assuming that the synchronous movement of oil film and water flow, and spilled oil concentration C unchanged along with time. Take the moving coordinate system. The diffusion equation of oil droplets can be simplified as follow:

$$\frac{\partial}{\partial z}\left(D_z \frac{\partial C}{\partial z}\right) + \frac{\partial}{\partial z}(VC) = 0 \tag{1}$$

where CO is the datum point concentration (mg/L);
when z=a, then C=CO;
when z→∞, then C=0.
where Dz is the vertical turbulent diffusion coefficient, taking for as a constant, then the solution of the equation (1) is:

$$C(z) = C_0 \exp\left[-\frac{V}{D_z}(z-a)\right] \tag{2}$$

This is the spilled oil concentration vertical distribution formula. Obviously, the concentration reduces exponentially.

Considering the relative motion slight between the oil droplets and water body, the oil droplets floating velocity V mentioned in the equation (2) can be calculated as follows:

$$V = \frac{(\rho_W - \rho_O)\, gd^2}{18\mu_W} \tag{3}$$

where d is the oil particle diameter (m); μw is the coefficient of viscosity of water (N S/m^2), ρ_0, ρ_w present density of oil and water respectively (kg/m^3). Some studies showed that the average diameter of oil droplets is 0.250 mm and the mean square error is 0.075 mm [30].

In addition, the vertical turbulent diffusion coefficient Dz can be calculated based on the following:

$$D_z = \alpha_z h u_* \tag{4}$$

where αz = empirical coefficient, and the suggested value is 0.067 [31]; h = water depth (m); u* = friction velocity.

It is necessary to input datum concentration C_0 into the equation (2) to predict the vertical concentration distribution of spilled oil. Therefore it is obvious that the characteristic parameter, datum concentration C_0, is the key limiting condition for the usage of the spilled oil vertical concentration distribution formula. However there are no appropriate field detection conditions and measurements. Obviously, it is impossible to acquire the datum concentration by actual measurement for early warning and forecasting of water quality. Therefore, it is the key technical problem to determine the datum point and its concentration, which decides whether the petroleum pollutants vertical concentration can get practical application in marine water quality warning and forecasting or not.

3 Determination of Datum Point Concentration

On the one hand, the spilled oil extension leads to expand the scope of pollution, which makes the oil film thinner. On the other hand, it can also increase the oil-water contact area, which will make more oil dissolve and emulsify into water body, strengthen the oil-water mixed degree, increase the oil concentration of water body, exacerbate the harm of the environmental protection goals [32]. Fay divided the oil membrane diffusion process into three stages, including gravity-inertia force stage, gravity-viscous force stage and surface tension-viscous force stage, according to the dominant driving force and resistance suffered.

First, when the oil film stays in the stage of surface tension-viscous force, it is thin and most of oil disperses into water [33]. Second, although there is still less proportion of oil film on the water body surfaces in this stage, considerate the most unfavorable risk effect. Therefore, it is assumed, the spilled oil could all dissolute and emulsify into the water and disperse evenly. Finally, set the reference point on the surface of water for solving the concentration C_0 easily.

The average spilled oil concentration of per unit area water column can be predicted from the following equation under the most unfavorable satiation.

$$\bar{C} = \frac{\rho_0 g D}{H} \tag{5}$$

where $\bar{C}$ = the average spilled oil concentration (mg/L); H = the depth of the water column (m); D = the oil film thickness (mm).

And then establish the response relationship of the vertical concentration and the surface concentration by integrating along the vertical direction to get the average concentration.

$$\bar{C} = \frac{\rho_0 g D}{H} \tag{6}$$

So the expression of C_0 can be showed by combining the equation (5) and (6):

$$C_0 = \frac{-K\rho_0 D}{e^{-KH} - 1} \tag{7}$$

Where $K = \frac{V}{D_z}$

Therefore, the concentration C_0 can be acquired by the equation (7) after inputting the corresponding data. Use the vertical distribution of concentration formula to get spilled oil concentration at different water depths.

4 Application

A petrochemical pier is located in some waters, and ships docking process may lead to oil spill accidents. The maximum credible oil spill quantity is 70t according to the source strength estimated method. Use the hydrodynamic model to simulate the hydrodynamic characteristics and take oil particle model to simulate the migration path of petroleum pollutants. Considering the risk prediction principle and adopting a conservative estimation, most spilled oil migrates into the related water column through the vertical diffusion when they stay in the stage of surface tension-viscous force. For the research waters, the coefficient of viscosity of water (μw) is $1.002\times10^{-3}N\cdot S/m^2$; the density of oil and water (ρ_0 and ρ_w) is 850 kg/m^3 and 1030 kg/m^3 respectively. And the diameter of oil particle is 250 μm. The vertical turbulent diffusion coefficient Dz is 0.002 m^2/s. Then, the surface concentration is available and it can be used to predict the characteristics of the vertical concentration distribution. The results are presented in Table 1 and Figure 4.

Table 1: Spilled Oil Concentration Vertical Distribution

vertical distance from water surface (m)	2	3	4	5	6	7
Concentration (mg/L)	22.1	1.7	0.13	0.01	0.0008	6.4×10^{-5}

5 Conclusions

In this paper, a mathematical calculation method for computing the surface spilled oil concentration is put forward. In a word, we establish the response relationship of the vertical concentration and the surface concentration under the most unfavorable situation by integrating the spilled oil concentration along the vertical direction. After a series of calculations, the surface concentration is available.

1. Based on the vertical concentration distribution formula of spilled oil, a calculation method of the datum point concentration, applied to the stage of surface tension-viscous force, is proposed. This method makes the usage of the vertical concentration distribution formula to analyze and forecast the possible vertical concentration distribution characteristics. Generally speaking, the characteristic parameter calculated by this mathematical method fills in the gap between theory and practice.

2. It is the application premise of this method that most of the oil has dissolved and emulsified into the water, applying for the stage of surface tension-viscous force mainly. And it can also apply for the second stage of spilled oil diffusion (stage of gravity-viscous force) after considering the safety and conservation of the risk impact assessment even though the results will be overestimated.
3. This method is not only suitable for oil spill accidents occuring in the sea, but also in the inland rivers, especially analyzing the impact from petroleum pollutants to the sensitive objects, and evaluating the accidental influence on the marine biological communities and water intakes which are located at different depths.

References

[1] Anderson C.M., LaBelle R.P., Update of comparative occurrence rates for offshore oil spills[J]. Spill Sci Technol Bull, 2000,6(5–6): 303-321
[2] Li Z, Kepkay P, Lee K, King T, Boufadel M.C., Venosa A.D., Effects of chemical dispersants andmineral fines on crude oil dispersion in a wave tank under breaking waves[J]. Mar Pollut Bull. 2007, 54(7): 983-993
[3] NRC (National Research Council) Oil in the sea III: inputs, fates and effects. National Academies Press, Washington, D.C. 2003
[4] Gong Y., Zhao X., Cai Z., O'Reilly S.E., Hao X., Zhao D.. A review of oil, dispersed oil and sedimentinteractions in the aquatic environment: influence on the fate, transport and remediation of oil spills[J], Mar Pollut Bull 2014, 79(1–2): 1 6-33
[5] Lytle D.A., Peckarsky B.L., Spatial and temporal impacts of a diesel fuel spill on stream invertebrates[J]. Freshw Biol, 2001, 46: 693-704
[6] Bandara U.C., Yapa P.D., Xie H., Fate and transport of oil in sediment laden marine waters[J]. J. Hydro-environ Res, 2011, 5(3): 145-156
[7] Bragg J.R., Owens E.H.. Clay-oil flocculation as a natural cleansing process following oil spills:part 1-studies of shoreline sediments and residues from past spills[C]. Proceedings of the 17th Arctic andMarine oilspill program (AMOP) Technical Seminar, Ottawa 1994, 1-23
[8] Wang H., Campbell J.R., Ditmars J.D., Computer Modeling of Oil Drift and Spreading in Delaware Bay[M]. Ocean Engineering, University of Delaware, 1975, 11-30.
[9] Lehr W.J., Cekirge H.M., Fraga R.J. et al. Empirical studies of the spreading of oil spills[J]. Oil and Petrochemical Pollution, 1984, 2(1): 7-11.
[10] Fingas M.F., The evaporation of oil spills: Development and implementation of new prediction methodology. Marine Environmental Modelling Seminar ' 98, 1998.
[11] YAN Z.Y., XIAO J.K., YIN P.H., Research on the characteristics of oil spill evaporation[J], Journal of Dalian Maritime University. 2000, 26(4): 21-24
[12] Robotham P.W.J, Gill R.A., Input, behavior, and fates of petroleum Hydrocarbons. The Fate and Effects of Oil in Fresh Water. London, Elsevier Science Publishers Ltd, 1989, 41-79.
[13] Oil in the sea III: Inputs, Fates, and Effects. Washington, D.C: The National Academies Press, 2003.
[14] Global Marine Oil Pollution Information Gateway (GMOPIG). What happens to oil in the water? Technical Information Paper, 2007.
[15] Minguez R., Abascal A.J., Castanedo S., Medina R., Stochastic Lagrangian trajectory model for drifting objects in the ocean[J]. Stoch Environ Res Risk Assess. 2012(26):1081-1093

[16] Mackay D., Paterson S., Trudel K., "A mathematical model of oil spill behavior," Environmental Protection Service, Fisheries and Environment Canada.
[17] Huang J.C., A review of the state-of-the-art of oil spill fate/behavior models[C]. in Proceedings of the Oil Spill Conference. Washington, 1983, 313-322
[18] Spaulding M.L., Oil spill trajectory and fate modeling: state-of-the art review[C]. in Proceedings of the 2nd International Oil Spill Research and Development Forum, London, UK, 1995.
[19] Mackay D, Buist I.A., Mascarenhas R., Paterson S., Oil spill processes and models[C]. Environment Canada Report EE-8, 1980.
[20] Mackay D., Paterson S., Trudel K., Mathematical model of oil spill behavior[C]. Environment Canada Report EE-7, 1980.
[21] Delvigne G., Sweeney C.E., Natural dispersion of oil[J], Oil Chem Pollut, 1988, 4(4): 281-310.
[22] Parsa R., Kolahdoozan M., Alavi Moghaddam M.R., Mid-depth oil concentration due to vertical oil dispersion in a regular wave field[J], Environ Fluid Mech 2016, 16(2): 335-346.
[23] Zhao W.Q., Jiang W. A study on the diffusion of droplet oil under water surface[J].Acta Scientiae Circumstantiae, 1990, 10(2): 173-182.
[24] Blumer M., Sass J., The west Falmouth oil spill, NTIS U.S. Department of Commerce, 1970.
[25] Moore S.F., Dwyer R.I., Katz A.M., A preliminary assessment of the environmental vulnerability of Machias Bay, Maine to oil supertankers. Report No. 162, Ralph M. Parsons Laboratory for Water Resources and Hydrodynamics. Mass. Inst. of Technol., Cambridge Mass, 1973
[26] IAHR IPD (International Association for Hydro-environmental engineering and research, innovationand professional development division), 2010. https://sites.google.com/site/oilspillsworkinggroup/.
[27] Fay J.A., The Spread of oil Slicks on a Calm Sea[M]. Oil on the Sea. Plenum Press, New York, 1969: 53-63.
[28] Wu X.D., Song J.M., Li X.G., Research on method for thickness of marine oil spill based on laboratory simulation and theoretical models[J]. The Ocean Engineering.2010.34(2):92-96.
[29] Yang Qingxiao, Xu Junying, Li Wensen. Research on the dissolution process of oil spill on sea. Marine Environmental Science. 1992, 9(11): 24-28.
[30] Cai Z.Q., Study on the Oil Spill into Partical and Adaptive Numerical Simulation of Ships at Sea[D]. Hubei: Wuhan Technology University, 2010.
[31] Zhang S.N., Explore the vertical diffusion regularity of nature river[J]. Water Resouces Protection, 1985.1.
[32] Cai J.Y., Han L.X., Study on prediction analysis of water pollution under emergency[M]. Nanjing: Hohai University Press, 2013
[33] Wang Y.H., Study on Spill Oil's Fate and Morphology on the Sea[D]. Liaoning: Dalian Maritime University, 2010.6.

Tao Ding*, Zhenni Guo, Chengxi Ying

Contamination Status and Potential Ecological Risks of Heavy Metal in Sediments from the Qiantang River

Abstract: Heavy metal sediment contamination can significantly damage the ecological water environment. In this paper, we collected superficial sediment samples at 26 different sampling sites, and then tested and evaluated their contamination status and potential ecological risks. The sediment analysis results showed that Cu, Zn, As, Cd, and Pb contents were excessive compared with local soil background levels. There is a low or moderate potential ecological risk level from heavy metal pollution in sediments in the study area; Cd is the element with the potential to cause the most ecological damage.

Keywords: component, heavy metal, ecological risk, river sediment, Qiantang River.

1 Introduction

With industrial development and population growth, environmental pollution caused by industrial wastewater and domestic sewage has become increasingly serious. Urban river pollution has gradually intensified. River sediment is an important component of the water ecological environment. On the one hand, sediments can absorb pollutants in water and relieve water pollution. However, sediment can also release pollutants back into the water upon external environmental changes and thus cause secondary water pollution [1,2]. Sediment is the main living place and food source for benthos. Pollutants in sediments could access terrestrial biota and human bodies through biological concentration and the food chain [3-5].

Urbanization, industrialization, and intensive agricultural development have increased heavy metal contents in river sediments [6]. Because of their toxicity, durability, and non-degradability [7], heavy metals have become key pollutants that significantly influence sediment quality [8,9] and cause sediment to be a secondary pollution source. Therefore, it is necessary to detect heavy metals in river sediments to evaluate heavy metal pollution levels and corresponding potential risks [10-15]. The results from this study will help develop strategies for river pollution control and the ecological dredging of rivers in the Zhejiang province, China, and will also provide reference information for similar areas.

***Corresponding author: Tao Ding,** College of Quality & Safety Engineering, China Jiliang University, Hangzhou, China, E-mail: dingtaohz@163.com
Zhenni Guo, Chengxi Ying, College of Quality & Safety Engineering, China Jiliang University, Hangzhou, China
This study was supported by Natural Science Foundation of Zhejiang Province (No. LY13E090005)

2 Materials and methods

2.1 Description of Study Area

In this paper, sampling sites were mainly in Hangzhou and Jiaxing along the Qiantang River. Table 1 shows the spatial distribution of sampling sites.

Table 1: Distribution of Sediment Sampling Sites

Sample Site	Longitude and latitude	
	Longitude	latitude
1	30°39′15″	120°31′49″
2	30°14′49″	120°07′40″
3	30°01′41″	120°01′03″
4	30°01′39″	120°00′28″
5	30°14′50″	120°07′37″
6	30°06′16″	120°12′58″
7	30°06′02″	120°12′49″
8	30°15′33″	120°03′44″
9	30°39′15″	120°31′49″
10	30°04′36″	120°12′03″
11	30°07′49″	120°12′44″
12	30°19′27″	120°08′57″
13	30°09′09″	120°13′52″
14	30°13′48″	120°08′11″
15	29°39′02″	119°04′42″
16	29°27′52″	119°15′28″
17	30°06′43″	120°11′13″
18	30°03′23″	119°57′38″
19	30°03′13″	119°56′52″
20	30°44′38″	121°01′49″
21	30°42′07″	121°16′09″
22	30°04′17″	120°11′07″
23	30°39′58″	120°30′32″
24	30°06′24″	120°14′07″
25	30°16′39″	120°10′54″
26	29°59′22″	119°41′8″

2.2 Sample Collection and Processing

Superficial sediment samples were collected with piston samplers (Fig. 1) and kept in sealed, labelled bags. The samples were naturally air-dried in a cool, ventilated laboratory. Impurities such as gravel and animal and plant residue were removed. Later, the samples were tiled on a piece of solid fiberboard and pressed with a glass rod. Samples were sifted by a 20-mesh sieve to remove oversized substances. Sifted samples were divided through a quartering sample method, and the samples were ground by an agate mortar until they could be sifted through a 100-mesh sieve. The processed samples were sealed for later tests.

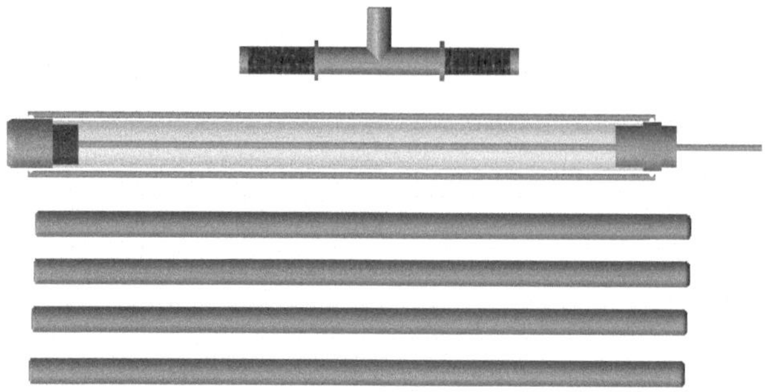

Figure 1: Sample analytical methods

2.3 Assessment of Potential Ecological Risk

The sample analysis process had 2 stages: sample pretreatment and heavy metal detection. The sample pretreatment steps were as follows: first, a weighed, dried sediment sample (0.4 g±0.0002 g) was put into a clean PTFE crucible, and 6 mL hydrochloric acid (HCl [36%]) was added on a hot plate at 100°C until only 2 mL of solution remained. When cooled, a combination of a certain volume of nitric acid (HNO_3 [69%]), hydrofluoric acid (HF [48%]), and 3 mL perchloric acid ($HClO_4$ [70%]) was added on a hot plate between 120 and 150°C until it was nearly dry. The digested sediment solution was diluted with deionized water and filtered quantitatively into a 50 mL volumetric flask. The heavy metal contents (Ni, Cu, Zn, As, Cd, and Pb) in the sediment samples after pretreatment were detected by ICP-MS.

2.4 Assessment of Potential Ecological Risk

Hakanson developed the Potential Ecological Risk Index [16] to evaluate damage to the environment from heavy metals in the sediment, and to assess heavy metal pollution. It integrates the concentrations of heavy metals, their toxic responses and ecological factors, and shows the combined effect of many types of heavy metals. The potential risk index can be calculated as follows:

$$C_f^i = C_h^i / C_n^i \tag{1}$$

$$E_r^i = T_r^i \times C_f^i \tag{2}$$

$$R = \sum_{i=1}^{n} E_r^i = \sum_{i=1}^{n} T_r^i \times C_f^i \tag{3}$$

where C_f^i is the contamination coefficient for a certain heavy metal, and C_h^i is the measured value of the heavy metal. C_n^i is a reference value for heavy metals. E_r^i is the potential ecological risk index for heavy metal pollution from one metal. T_r^i is the toxic response factor of heavy metals that indicates the hazards of heavy metals on human and aquatic ecosystems, and reflects the levels of heavy metal toxicity and ecological sensitivity to heavy metal pollution. RI represents the potential ecological risk index of multiple heavy metals.

In this study, we adopted the environmental background levels of heavy metals from Wang et al. [17] to evaluate the pollution level. Table 2 lists the local background value and the toxic response factor of heavy metals in sediments.

Table 2: The Local Background Value and the Toxic Response Factor of Heavy Metals in Sediments [16,17].

Metals	Ni	Cu	Zn	As	Cd	Pb
C_n^i	41.10	40.80	110.00	10.00	0.21	38.20
T_r^i	1	10	5	5	30	5

3 Results and discussion

3.1 Concentrations and Distribution of Heavy Metals in Sediments

Based on Equation (1), heavy metal concentrations in sediment surfaces (C_h^i) should be measured. Table 3 presents the concentrations of Cu, Zn, Pb, Cd, AS, and Ni in surface sediments. Ni concentrations were lower than the background levels at most sampling sites (except No.18). The Fig. 2 for Cu was also lower than the background levels at most sampling sites (except across 4 sites), but was nearly 10 times higher than the background level at site No.17. On the other hand, the concentrations of other metals (Pb, Cd, As, and Zn) were higher than the background levels at more than half of sampling sites, especially for Cd. The standard exceedance rates of Cu, Zn, As, Cd, and Pb were 4%, 15%, 27%, 50%, 81%, and 58% respectively. The maximum concentrations of Cu, Zn, As, Cd, and Pb were 9.90, 3.86, 2.23, 15.74, and 3.43 times greater than the background values, respectively, while the average concentrations of Cu, Zn, As, Cd, and Pb were 1.09, 1.28, 1.07, 3.62, and 1.17 times greater than the background values, respectively. Of all the metals, pollution from Cd is the most serious. The pollution levels decrease in the order Cd>Zn>Pb>Cu>As.

Table 3: The Concentrations (mg kg^{-1}) of Heavy Metals in Surface Sediments.

Sampling site	Ni	Cu	Zn	As	Cd	Pb
1	34.02	23.20	62.71	0.01	0.27	32.14
2	37.72	34.52	151.46	10.30	0.91	40.12
3	20.77	29.70	54.21	9.79	1.12	81.80
4	22.66	34.85	105.37	9.55	1.15	87.17
5	17.61	51.61	861.55	10.03	0.75	48.13
6	26.84	35.32	77.72	10.51	1.46	64.46
7	24.02	89.31	256.49	10.66	1.42	50.79
8	34.04	24.49	85.59	14.59	0.28	29.83
9	25.48	24.67	118.72	17.22	0.48	35.29
10	20.90	19.58	65.46	8.42	0.37	28.22
11	21.71	26.00	80.93	13.31	1.21	60.20
12	32.04	25.72	56.95	8.81	0.13	28.36
13	31.96	33.83	107.68	9.20	0.96	67.80
14	27.45	20.50	59.76	9.61	0.20	43.81
15	36.98	78.79	355.70	13.41	2.94	131.01
16	26.03	16.40	78.20	9.96	0.20	23.08
17	25.90	403.75	425.04	16.60	3.31	55.05
18	66.29	29.49	80.89	22.35	0.16	19.96
19	35.98	21.95	94.32	7.05	0.27	26.30
20	29.14	22.39	55.71	6.59	0.25	29.49
21	33.23	24.71	62.46	5.83	0.35	33.48
22	27.12	26.79	123.52	16.73	0.53	34.70
23	13.14	12.84	75.45	8.85	0.37	37.38
24	14.83	8.90	49.56	14.85	0.13	16.43
25	22.56	17.16	50.26	3.81	0.28	30.45
26	30.37	17.81	77.04	11.06	0.25	23.64
mean	28.41	44.40	141.26	10.74	0.76	44.58
standard deviation	1.21	9.07	1.55	4.52	114.05	3.28

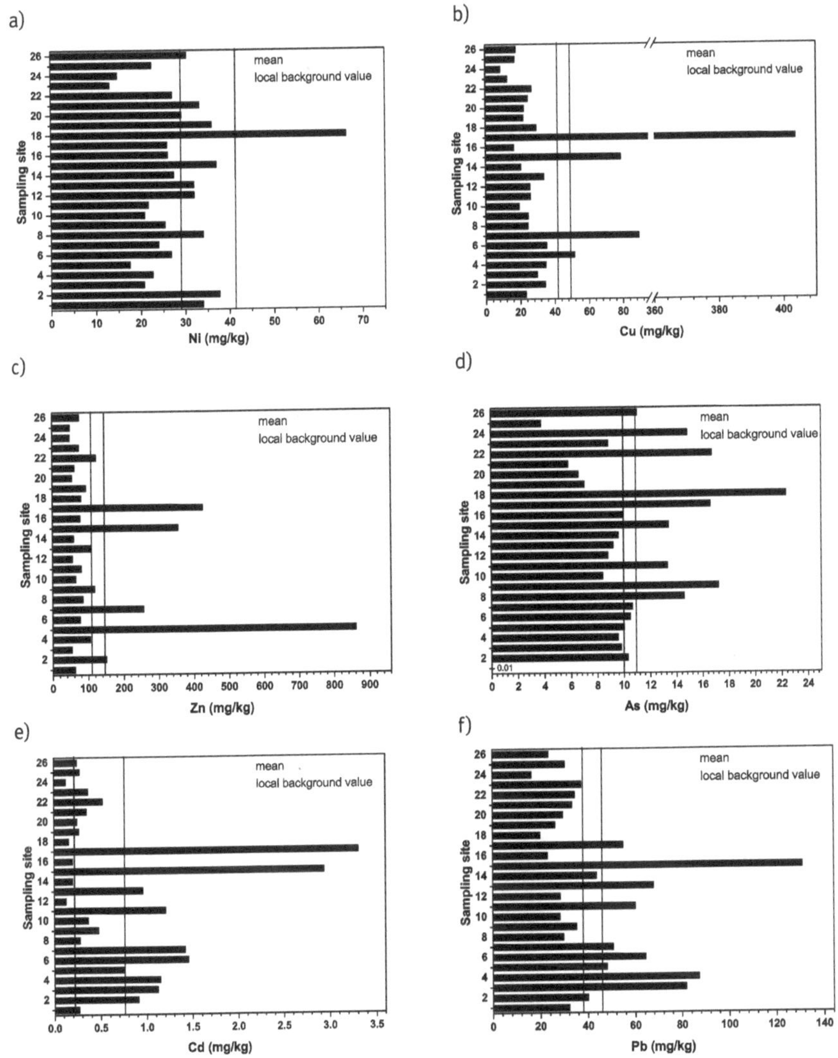

Figure 2: Concentrations histograms of Ni(a), Cu(b), Zn(c), As(d), Cd(e), and Pb(f) in surface sediments (mg kg^{-1})

The variation coefficient was used to analyse the spatial distribution of heavy metal content in the surface sediments of the study areas. It can be calculated as follows:

$$CV = Sn/Ln \quad (4)$$

Where, for a certain heavy metal, CV is the variation coefficient, Sn is the standard deviation of the 26 sampling sites, and Ln is the mean value of the 26 sampling sites.

Table 4 lists the variation coefficients of heavy metals in sediments. The variation coefficient decreased in the order of Cd>As>Cu>Pb>Ni>Zn. With the exception of Cd, all heavy metals had small variation coefficients. This indicates the spatial distribution of each heavy metal was not uniform, mainly because the rivers suffer from different pollution sources.

Table 4: Variation Coefficients of Heavy Metals in Sediments

Ni	Cu	Zn	As	Cd	Pb
0.04	0.20	0.01	0.42	150.07	0.07

Fig. 3 presents the distribution of heavy metals in the sediments of the study areas, and RI (Table 5) was used to present the pollution level. Based on the above analysis, we divided this part of the river into three ecological function areas: upstream, midstream, and downstream. It is necessary to take steps to control heavy metals upstream (Cu, Zn, As, Cd, and Pb) and midstream (Ni, Cu, Zn, As, Cd, and Pb).

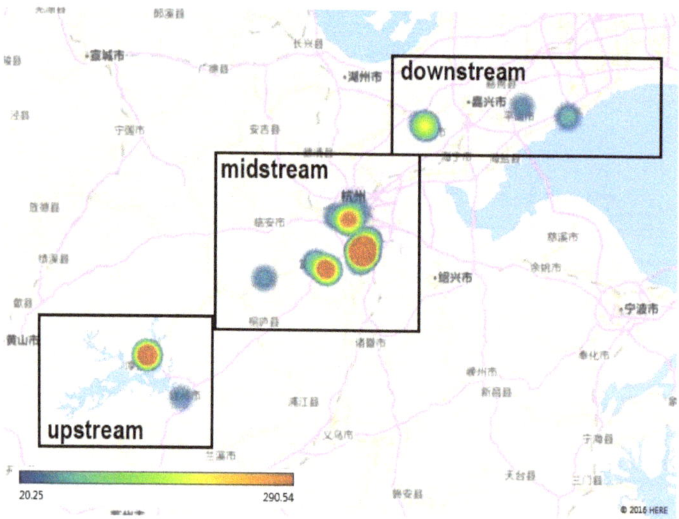

Figure 3: Distribution of heavy metals in sediment (RI)

3.2 Potential Ecological Risk of Heavy Metals in Sediments

Table 5 presents the result of the calculations of the single potential ecological risk index (E_r^i) and the integrated potential ecological risk index (RI) based on equations (2) and (3). It also lists the values of E_r^i and RI at each sampling site. The relation between E^i_r, RI, and degree are listed below (Table 6), and the potential ecological risk assessments in the study areas are shown in Fig. 4.

Table 5: The Potential Ecological Risk Assessment for Heavy Metals in Sediments

Sampling sites	E^i_r						RI
	Ni	Cu	Zn	As	Cd	Pb	
1	4.14	2.84	0.57	0.01	38.64	4.21	50.42
2	4.59	4.23	1.38	10.30	130.25	5.25	156.00
3	2.53	3.64	0.49	9.79	160.55	10.71	187.70
4	2.76	4.27	0.96	9.55	164.41	11.41	193.36
5	2.14	6.32	7.83	10.03	106.96	6.30	139.59
6	3.26	4.33	0.71	10.51	208.86	8.44	236.11
7	2.92	10.94	2.33	10.66	203.26	6.65	236.77
8	4.14	3.00	0.78	14.59	40.45	3.90	66.86
9	3.10	3.02	1.08	17.22	68.66	4.62	97.70
10	2.54	2.40	0.60	8.42	53.51	3.69	71.16
11	2.64	3.19	0.74	13.31	173.11	7.88	200.86
12	3.90	3.15	0.52	8.81	18.79	3.71	38.88
13	3.89	4.15	0.98	9.20	137.81	8.87	164.89
14	3.34	2.51	0.54	9.61	29.12	5.73	50.86
15	4.50	9.66	3.23	13.41	420.39	17.15	468.34
16	3.17	2.01	0.71	9.96	28.03	3.02	46.91
17	3.15	49.48	3.86	16.60	472.20	7.21	552.51
18	8.06	3.61	0.74	22.35	22.90	2.61	60.28
19	4.38	2.69	0.86	7.05	38.30	3.44	56.71
20	3.55	2.74	0.51	6.59	35.59	3.86	52.83
21	4.04	3.03	0.57	5.83	50.43	4.38	68.28
22	3.30	3.28	1.12	16.73	75.68	4.54	104.66
23	1.60	1.57	0.69	8.85	53.15	4.89	70.76
24	1.80	1.09	0.45	14.85	18.17	2.15	38.51
25	2.74	2.10	0.46	3.81	40.08	3.99	53.18
26	3.69	2.18	0.70	11.06	35.05	3.09	55.78

Table 6: The Relation Between E^i_r, RI, and Degree [16]

E^i_r and degree		RI and degree	
$E^i_r<40$	low	RI<150	low
$40 \le E^i_r<80$	moderate	150≤RI<300	moderate
$80 \le E^i_r<160$	higher	300≤RI<600	severe
$160 \le E^i_r<320$	high	RI>600	serious
$E^i_r \ge 320$	serious		

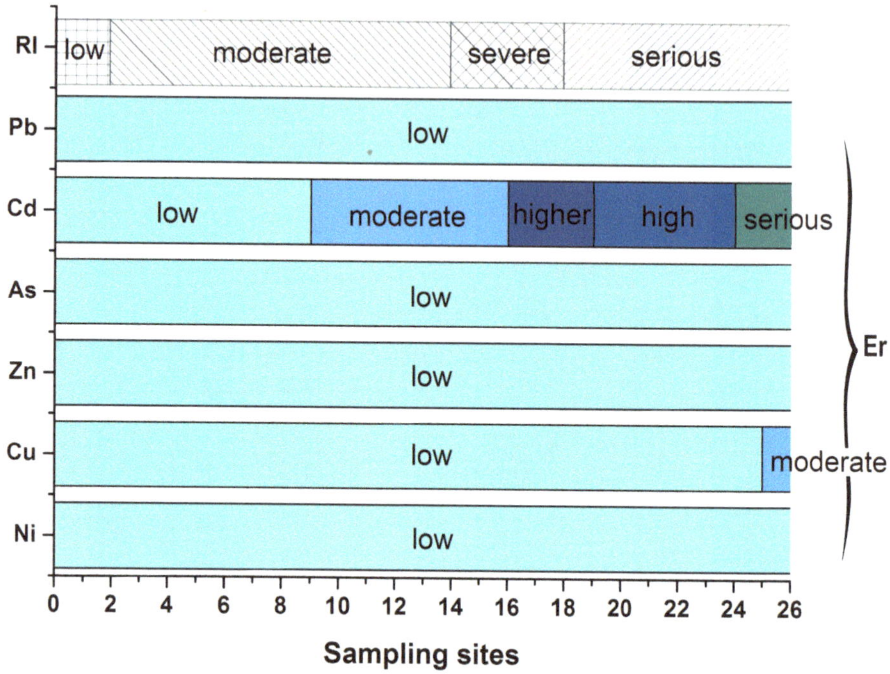

Figure 4: The potential ecological risk assessment for heavy metals in the study areas

Most potential ecological risk indexes (E^i_r) for Cd exceeded 40 (Fig. 3), and nearly half of them exceeded 80 and fell into the higher risk category. The E^i_r values for Ni, Zn, As, and Pb were less than 40, and indicated low risk. The E^i_r of Cu ranged from 1 to 10 (except one in 49.98), which indicated low risk. The potential ecological risk values (E^i_r) decreased in the order of Cd>As>Pb>Cu>Ni>Zn. RI values ranges from 30 to 600, and around half of them were higher than 150, which indicated that the integrated potential ecological risk at nearly 13 sampling sites was moderate. The average contribution proportions of E^i_r of Ni, Cu, Zn, As, Cd, and Pb to RI were 2.55%, 4.02%, 0.95%, 7.93%, 80.24%, and 4.31%, respectively. Of these, Cd contributed the most. These results show that Cd is the element with the potential to cause most ecological damage.

4 Conclusions

1. According to heavy metal distribution in sediments, we divided this part of the river into three ecological function areas: upstream, midstream, and downstream. It is necessary to take steps to control heavy metals in the upstream (Cu, Zn, As, Cd, and Pb) and midstream (Ni, Cu, Zn, As, Cd, and Pb) portions of the Qiantang River, China.

2. Comparison shows that pollution from Cd is more severe than the pollution from other metals, with an average concentration 3.62 times higher than the background levels. The risk from Cu is slight/medium, while the risks from Ni, Zn is low.

Acknowledgment: This study was supported by Natural Science Foundation of Zhejiang Province (No. LY13E090005).

References

[1] Vallee B.L., Ulmer D.D., Biochemical Effects of Mercury, Cadmium, and Lead, Annual Review of Biochemistry, 1972, 41, 91-128.
[2] Wang S., Jia Y., Wang S., Xin W., He W., Zhao Z., Liu B., Fractionation of heavy metals in shallow marine sediments from Jinzhou Bay, China, Journal of Environmental Sciences, 2010, 22, 23-31.
[3] Ajmal M., Raziuddin, Khan A.U., Heavy metals in water, sediments, fish and plants of river Hindon, U.P., India, Hydrobiologia, 1987, 148 151-157.
[4] Papagiannis I., Kagalou I., Leonardos J., Petridis D., Kalfakakou V.,, Copper and zinc in four freshwater fish species from Lake Pamvotis (Greece), Environment International, 2004, 30, 357-362.
[5] Ebrahimpour M., Mushrifah I., Seasonal variation of cadmium, copper, and lead concentrations in fish from a freshwater lake., Biological Trace Element Research, 2010, 138, 190-201.
[6] Hadi H.E., Assessment of Heavy Metals Soil Contamination in the Abandoned Pb-mine of Zaida (Morocco), European Journal of Scientific Research, 2015, 129, 167-178.
[7] Fu W., Meng F., Wang Z., Wang Q., Li Y., Zhou Y., Cheng F., Heavy metals in the intertidal sediments and two marine bivalves along the Beibu Bay: Contamination status and bioaccumulation, Acta Scientiae Circumstantiae, 2013, 33, 1401-1409.
[8] Bibi M.H., Ahmed F., Ishiga H., Assessment of metal concentrations in lake sediments of southwest Japan based on sediment quality guidelines, Environmental Geology, 2007, 52, 625-639.
[9] Yu L., Zhi-ming Y., Xiu-xian S., Application of Principal Component Analysis (PCA)for the Estimation of Source of Heavy Metal Contamination in Marine Sediments, ENVIRONMENTAL SCIENCE, 2006, 27, 137-141.
[10] van den Berg GA., Meijers GG., van der Heijdt LM., Zwolsman JJ., Dredging-related mobilisation of trace metals: A case study in The Netherlands, Water Research,2001, 35, 1979-1986.
[11] Weber P., Behr E.R., Knorr C.D.L., Vendruscolo D.S., Flores E.M.M., Ssler V.L., Baldisserotto B., Metals in the water, sediment, and tissues of two fish species from different trophic levels in a subtropical Brazilian river, Microchemical Journal, 2013, 106, 61-66.
[12] Hakanson L., An ecological risk index for aquatic pollution. A sedimentological approach, Water Research, 1980, 14, 975-1001.
[13] Nouri M., Haddioui A.E.M., Assessment of metals contamination and ecological risk in ait Ammar abandoned iron mine soil, Morocco, Ekologia, 2016, 35, 32-49.
[14] Wang F., Huang Y., Wang X., Gao Z., Faxin Y.U., Feng X.U., Bao Q., Ying H.U., Qiao M., Jin S., Ecological risk assessment of heavy metals in surrounding soils of tungsten ores: Comparison of different evaluation methods, Environmental Chemistry, 2015.
[15] Solgi E., Ahmadi F.Y., Ecological risk assessment of cadmium and lead in urban and forest park soils in Asadabad City, Iran., Journal of School of Public Health & Institute of Public Health Research,2015.

[16] Hakanson L., An ecological risk index for aquatic pollution control.a sedimentological approach, Water Research, 1980, 14, 975-1001.
[17] Wang Q.H., Dong Y.X., Zhou G.H., Zheng W., Soil Geochemical Baseline and Environmental Background Values of Agricultural Regions in Zhejiang Province., Journal of Ecology & Rural Environment, 2007, 23, 81-88.

Jingxin Su*, Zhiping Wang, Tianjing Wang, Aizhen Ji, Liguo Hou, Xiaomei Guo

Influence of Atmosphere Environment on Aircraft Heat Exchanger Fouling Growth Characteristics

Abstract: During domestic civil aviation operation, problems of fouling on the cooling surface of the aircraft heat exchanger have been more serious than ever due to air pollution. By using analytical methods including energy dispersive spectroscopy (EDS), energy dispersive X-ray (EDX) and ion chromatography respectively, sources of aircraft air-conditioning heat exchanger dirt are verified to be mainly air pollutants and soil particles. And by observing dirt growth characteristics under different surface states, it's found that when there are large particles such as catkins on the surface of samples, the rate of fouling is increased for the reason that large particles will augment adsorption sites of nucleation. It turns out that dirt nuclei selectively concentrate in the vicinity of large particles instead of generating randomly. While oil and residual oily cleaning agents can inhibit the growth of fouling by isolating test pieces from atmosphere.

Keywords: aircraft heat exchanger, thin film, crystallization fouling, nucleation sites.

1 Introduction

Heat exchanger is an important component of the aircraft air-conditioning system, whose reliability plays a decisive role for the reliability of the whole aircraft environmental control system. There are different degrees of fouling problems for most of the heat exchangers, wherein its fouling problem on cooling surface has been a widespread problem during domestic civil aircraft operation. The air conditioning system failure rate of Boeing 737, the main trunk of most airlines, is relatively high, especially in hot summer. Faults often appear one after another, which not only influence line maintenance, aircraft release and other produces, but also affect the operating efficiency and maintenance costs of the whole aviation business [1]. Once the working efficiency of heat exchanger effect is so poor that cold airflow temperature reaches any protection temperature, thermal protection

***Corresponding author: Jingxin Su,** Tianjin Key Laboratory for Civil Aircraft Airworthiness and Maintenance, Civil Aviation University of China, Tianjin, China, E-mail: sujingxin@139.com
Zhiping Wang, Aizhen Ji, Tianjin Key Laboratory for Civil Aircraft Airworthiness and Maintenance, Civil Aviation University of China, Tianjin, China
Tianjing Wang, Liguo Hou, Xiaomei Guo, Aircraft Maintenance and Engineering Corporation, Beijing Capital International Airport, Beijing, China

system will automatically shut off assembly valve and the air conditioning system will stop working [2].

Current research focuses more on fault repair, rather than develop into the root causes of failure. Therefore, understanding the mechanism of the formation and growth characteristics of dirt, determining cleaning cycle and optimizing cleaning process according to the characteristics of different growth stages of dirt become more important. From hydro-dynamics and physicochemical conditions two viewpoints, Yiantsions [3] analyzed the deposition process of micron particles in the heat exchanger surface, whose experimental results reveal the interaction between gravity and other physical and chemical factors. By assuming crystallization dirt and particles dirt exist independently, crystallization occurs simultaneously on the heat exchanger surface and the particle surface as well. Xu Zhiming [4] used the area ratio to divide calcium carbonate deposition into two parts: Part of the calcium carbonates are deposited on the wall, forming a crystallization fouling; another part of the calcium carbonate particles deposit on the surface of particles dirt, joining deposition process of particles dirt. Zhang Zhongbin [5] studied the effects of heat transfer surface characteristics and working fluid flow characteristics on fouling induction period by the method of comparative experiments, results showed that fouling induction period of the heat transfer was extended with the reduction of material surface adhesion and the enhance of field disturbance. Roya Sheikholeslami [6] demonstrates that other non-crystallizing species, if present, may affect the surface energy of a crystal and alter the Gibbs free energy for homogeneous nucleation and other crystallizing species may also act as a nucleus initiating nonhomogeneous nucleation with a lower Gibbs free energy and promoting nucleation.

In order to observe the growth characteristics of crystallization fouling, Won Tae Kim [7] used microscopy to observe the growth of $CaCO_3$ crystallization fouling, whose experimental device was designed to observe the growth process of dirt in a sealed system, which cannot be used very well to study the growth of crystallization dirt under open system.

In this paper, analytical methods including EDS, EDX and ion chromatography are used to verify the source and main ingredients of aircraft air-conditioning heat exchanger. Using the analytical results obtained, with reference to thin film research methods, we study the crystallization fouling growth characteristics of aerosol whose main soluble ions are Cl^-, NH_4^+, NO_3^-, SO_4^{2-} and Na^+ under humid atmosphere. Soluble salt mixed solution was prepared to form a thin film on the specimen surface and placed in the simulated work environment. The growth of the dirt was under open systems, using microscopic observation for the growth process of air side fouling.

2 Experimental Methods

2.1 Aircraft Air-conditioning Heat Exchangers Pollution Analysis

According to their present status, air pollutants can be classified as aerosols pollutants and gaseous status pollutants. Aerosols pollutants mainly include dust, mist, fly ash and suspended solids; Gaseous pollutants include sulfur oxides, nitrogen oxides, carbon oxides and hydrocarbons.

The atmospheric aerosol, referring to the multi-phase systems composed of the atmosphere and the solids and liquid particles suspended in it, which includes Total Suspended Particles (TSP), particles with an aerodynamic diameter less than or equal to 100 μm suspended in the atmosphere and water-soluble ions. Water-soluble ionic components SO_4^{2-}, NO_3^-, Cl^-, NH_4^+, K^+, Na^+, Ca^{2+} and Mg^{2+} account for a large proportion in atmospheric aerosols. In recent years, Chinese environmental scientists have made a lot of progress in the study of atmospheric particulates [8]. It summarizes the most common source of part ions in aerosols as shown in Table 1.

Table 1: Common Source of Part Ions in Aerosols.

Ions	Common Source
SO_4^{2-}	Combustion of coal, oil and other fossil fuels
NO_3^-	Oxidation of NO_X produced by combustion of fossil fuel and natural gas
Cl^-	Combustion of waste garbage, industrial production
NH_4^+	Farming and animal husbandry, such as rotting plants, animal wastes and fertilizer use
K^+	Emission of soil or human life combustion
Na^+	Form of NaCl
Ca^{2+}	Soil dust and construction
Mg^{2+}	Soil dust

Due to the large surface area, it is easy for small particles to absorb plenty of organics on its surface [9]. Therefore, the organic pollutants are also an important component of air pollution. Particulate organic matters are composed of hundreds of organic compounds, with widely varied chemical properties and thermodynamic properties, which bring a lot of difficulties for the study [10]. But through the accumulation of research, environmental scientists achieved certain results on the chemical composition and characteristics of particulate organic matter. Organic pollutants in the atmosphere mainly come from direct emissions, including incomplete combustion of fossil fuels and biomass, source plant emissions and natural fire, while some of it is the secondary source generated by chemical reaction of gaseous organic compounds [11].

2.2 Dirt Collection of Airborne Heat Exchangers

After theoretical analysis of the heat exchanger fouling sources, we understand the possible composition of dirt. And to further verification, the analysis of the dirt collection through the heat exchanger is needed. Direct collection from the heat exchanger surface not only can destroy the heat exchanger, but also the amount of dirt obtained is so small that it is not easy to analyze its composition. Therefore, in this paper, we collected dirt from heat exchanger by indirect methods and analyzed it.

The performance of contaminated heat exchanger will degrade which can easily lead to failure of the air conditioning system, thus the heat exchanger is periodically cleaned. During the cleaning process, the cleaning agent will dissolve part of the dirt which reduces the binding forces between dirt, resulting in some dirt off. This paper analyzes the insoluble dirt off particles during the cleaning process of heat exchanger. Separating insoluble dirt particles from the cleaning agent by using high-speed centrifuge, after which dirt particles were put in a drybox under 80°C for 24 h to obtain powdery particles of dirt. Analysis of the powder can be done by means of fluorescence spectroscopy to obtain elemental composition and mass fraction of insoluble dirt.

In addition, cut the collected waste heat exchanger to observe the cold road fin within it. Using scanning electron microscopy to observe the dirt on the surface of fin and X-ray spectroscopy to analyze the elemental composition, elemental composition of dirt on fin surface can then be obtained.

The results obtained by the two experimental programs above are just the elemental composition of dirt, but not the direct expression of dirt components. The working principles of aircraft heat exchanger and outdoor heat exchanger are the same, of which the main sources of dirt are aerosols pollutants and particulates. So in order to be more intuitive for the understanding of the composition of dirt, this experiment has collected dirt on the heat exchanger after perennial work outdoors. Immersed in deionized purified water for 240 h, separate the insoluble dirt particles by centrifugation and analyze the solution by ion chromatographic method to detect the soluble ions in it. Dirt sources, analytical procedures and the applicability of methods are all in Table 2.

Table 2: Part of Soluble Ions in Aerosols.

Sources of Dirt Collection	Analytical Procedures	Applicability of Methods
Insoluble particles obtained by centrifugation of cleaning agent	EDX	Analysis of elemental composition and percentage of weight of the particles
Dirt on the surface of heat exchanger	EDS	Analysis of elemental composition and percentage of weight of the particles
Outdoor air conditioning heat exchanger	Ion Chromatography	Analysis of soluble anion and cation in solution

2.3 Crystallization Growth Characteristics under Thin Film Environment

The experiments use a self-designed thin film equipment to study crystallization growth characteristics, as shown in Fig. 1. Wherein Fig. 1a is a principle diagram of thin film device and Fig. 1b is a front view of the thin film device. Thin film device consists of thin liquid film work systems, thickness measuring system and observation system. Thin film measuring device [12] is mainly composed by ohmmeter and screw micrometer equipped with fine platinum needle.

When measuring the thickness of thin film, fix the screw micrometer stuck with platinum needle directly above the test piece and adjust the micrometer screw slowly and carefully to enable platinum needle close to the specimen surface. When platinum needle tip gets in contact with the electrode surface, the ohm resistance shown in the table jumps from the infinite to a fixed value, then stop the adjustment of screw micrometer immediately and read the screw micrometer scale, denoted by h_1. Then getting the platinum needle back away from the test piece surface, the deionized water flowed through the inlet sweep over the whole specimen. Now adjust the screw micrometer again to get the platinum needle close to the surface of the solution until they are in contact, read screw micrometer scale h_2, so that the thickness of the film may be referred to as $\triangle h=h_2-h_1$. The operating temperature of aircraft air-conditioning heat exchanger cold road is usually stable at 80°C - 120°C, over which it will start self-protection and shut down the system, so this experiment controls the temperature of thin film device at 80°C by heating station. Close pure water flows after that deionized pure water sweep over the specimen, with the thin film device at 80°C, pure water will evaporate which decreases thin film thickness. At this time since solution is deionized pure water, there will be no precipitation of dirt at this stage called preparation of thin film. Continuously measuring the value h_2, open the peristaltic pump when $\triangle h$ is less than 300 μm, letting the dirt solution flow into the thin film in accordance with a certain velocity. By controlling the flow rate of the peristaltic pump, the thin film can be stabilized at 200 μm - 300 μm. After repeated experiments, we found that when the inlet flow rate is controlled between 320 μL and 350 μL, the thin liquid film thickness can be controlled at 200 μm - 300 μm. Thereafter, the thickness measuring system can be replaced by observation system, photographing at a certain time to observe the growth of fouling.

In order to ensure that the dirt precipitation can be accelerated, in this study, mixed solution of NH_4Cl, Na_2SO_4 and $NaNO_3$ with 80% molar concentration respectively are prepared as the analog solution. Under actual operating conditions, the surface of the heat exchanger does not always maintain the clean state, but there will be a wide variety of dirt adsorbing on the surface, so the surface of the test sample was divided roughly into four categories for analysis: without any treatment, that is, clean state; catkins attached to the surface which characterize the state of large particles adsorption on the heat exchanger surface; attached to aviation kerosene, showing the state of a heat exchanger attached with organic pollutants in the atmosphere;

with cleaning agent, that is when the heat exchanger wasn't washed clean. Therefore the experiments were divided into four groups wherein the sample surface of the first group received no treatment. And during thin film preparation process of second group, catkins will be spread evenly in thin liquid film surface. In front of the thin film preparation process of the third and fourth group, aviation kerosene and Turco3878LF-NC cleaning agent coated evenly on specimen surface. After that, start thin film preparation until it achieves the demanded degree, then open peristaltic pump, dirt solution slowly flowing on thin liquid film surface by a certain velocity.

(a) Thin film device operating principle (b)Thin film device front view

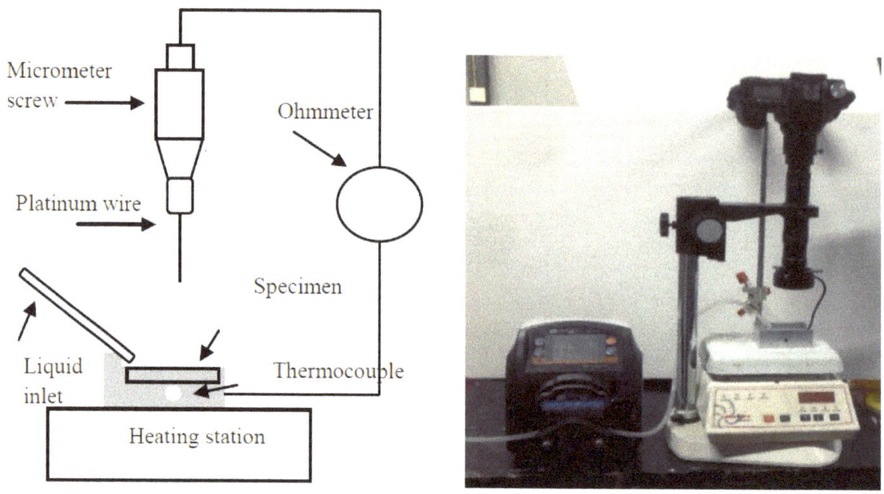

Figure 1: Designed thin film equipment

3 Results

3.1 Main Ingredients of Aircraft Air-conditioning Heat Exchangers

By taking a small portion of the fin from the abandoned aircraft heat exchanger surfaces, we use scanning electron microscopy X-ray energy dispersive spectroscopy to analyze surface dirt micro-element composition. Analysis results are shown in Fig. 2 and Table 3. Collecting waste liquid of heat exchanger after cleaning from an aircraft maintenance company, insoluble dirt is obtained by centrifugation. Elemental composition analysis can be done by EDX, the results are shown in Table 3.

As it can be seen from Table 3, major elements in dirt are Na, Mg, Si, K, Ca, Fe, Cu and O, which are common elements in the atmosphere or soil particles. The insoluble particles obtained by centrifugation can be observed by a laser microscope, shown in Figure 3. Insoluble particles in dirt are mainly various colors of minerals, of which main source are soil particles, further validating the foregoing analysis about dirt sources of heat exchanger. And through the ion chromatographic analysis about outdoor heat exchanger, main soluble ions: Na^+, NH_4^+, K^+, Mg^{2+}, Ca^{2+}, Cl^-, NO_2^-, Br^-, NO_3^- and SO_4^{2-}, which is very close to the elemental analysis results of aircraft air-conditioning heat exchanger. In conclusion, the source of aircraft air-conditioning heat exchanger dirt are mainly air pollutants and soil particles.

Figure 2: SEM laser spectroscopy bit

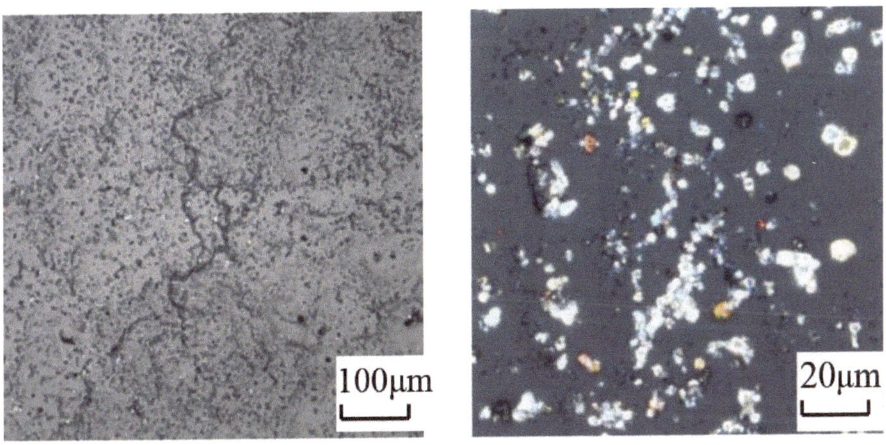

Figure 3: Insoluble dirt particle morphology under a laser microscope

Table 3: Analysis of Results of Dirt Sample Elements

Element	**O**	**Na**	**Mg**	**Al**	**Si**	**P**	**S**	**K**
EDS	19.93	2.24	2.98	42.61	3.87	2.31	1.22	1.28
EDX	-	7.283	2.288	11.96	27.17	3.87	-	4.385
Element	**C**	**Ti**	**Cl**	**Ni**	**Co**	**Ca**	**Fe**	**Cu**
EDS	-	-	-	-	-	1.37	20.19	1.82
EDX	24.8	1.36	0.665	0.403	0.0226	2.863	12.2	0.122

3.2 Weight Gain Curve and Surface Topography of Crystallization

Fig. 4 is the crystallization fouling weight gain curve under different surface states, which shows that the fastest growing of crystallization fouling is when large dirt particles such as catkins adhesion on the surfaces whereas dirt growth has been somewhat suppressed when there are oil or residual cleaning agents on the surface. This is because the presence of oil and oil-based cleaning agents specimen surface prevent the thin film from direct contact with the test piece so that after the formation of dirt nucleation, it cannot be adsorbed on the test piece surface but floating on oil surface, only large particles of dirt can be deposited and adsorbed onto the surface of the test piece. Dirt nucleation floating on the oil surface are easily taken away by flowing thin film, it is not easy to complete this series of nucleation and growth process in the test piece surface.

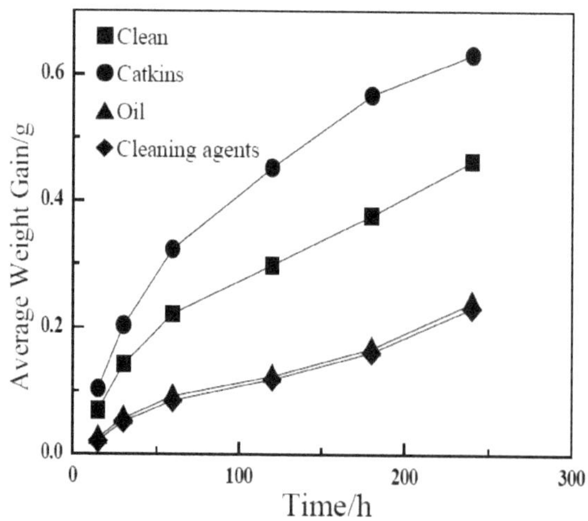

Figure 4: Crystallization fouling weight gain curve under different surface states

To further study the differences of crystallization fouling, the sample surface topography at different times are taken during the testing process as shown in Fig. 5-8. Fig. 5 shows the growth process of dirt on the surface of flat specimen polished using the #1200 sandpaper. Some scholars have studied the influence of different surface roughness on fouling growth performance, founding that dirt adsorption on the surface of the sample exhibits random results [13].

Therefore, this paper does not consider the effect of roughness on the growth of fouling. During the experiment, on the thin film a lot of small bubbles appeared randomly, as shown in Fig. 5b. As the bubbles escaped, local concentration increased and salt crystal nucleation was formed here. Deposited on the test piece surface, part of dirt nucleation was taken away by the flowing thin film, while others which were adsorbed on the surface of the sample continued to grow, as shown in Fig. 5c. The form of deposited dirt is like dendrite and the staggered growth gives them a strong binding. Bubble randomness also illustrates absorption in the dirt is with certain randomness.

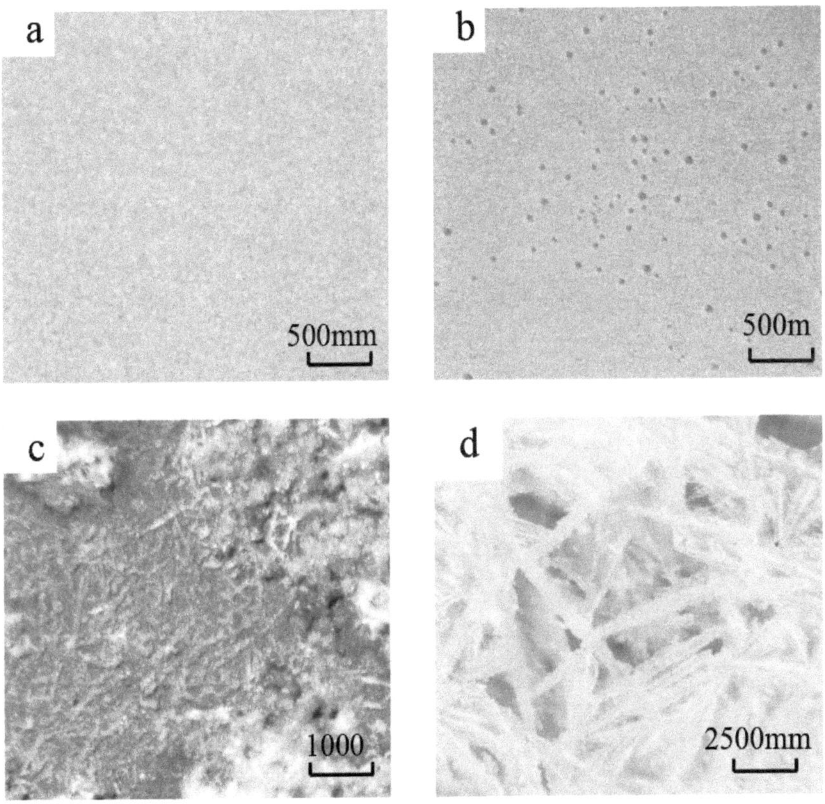

Figure 5: Topography of sample with clean surface over time, a: 0 h; b: 0.5 h; c: 2 h; d: 4 h

Fig. 6 illustrates dirt growth characteristics when catkins attached to the surface. As can be seen from Fig. 6b, the bubbles don't occur randomly, but selectively concentrated in the vicinity of catkins. This is because when there are catkins and other solids in a thin film, the surface energy is reduced, so that air bubbles in the thin film will be attached and accumulated on the surface of catkins. Large bubbles formed by the accumulation of small bubbles can easily escape from the thin film, increasing local concentration. Dirt nucleation does not need to be on the sample surface, but grows in the surface of catkins. Besides, additional crystal nuclei can significantly improve crystallization rate effects [14-15]. Meanwhile, the nuclei adsorbed on catkins are not likely to be taken away by flowing thin liquid film, so dirt growth was relatively rapid. And after that, catkins were firmly adsorbed on the surface of the test piece, easy to form stubborn dirt, as shown in Fig. 6c. As can be seen from Figure 6d, the presence of catkins made the binding of dirt more compact, on the one hand increased the absorption point of dirt nucleation, and on the other side increased the binding ability between dirt.

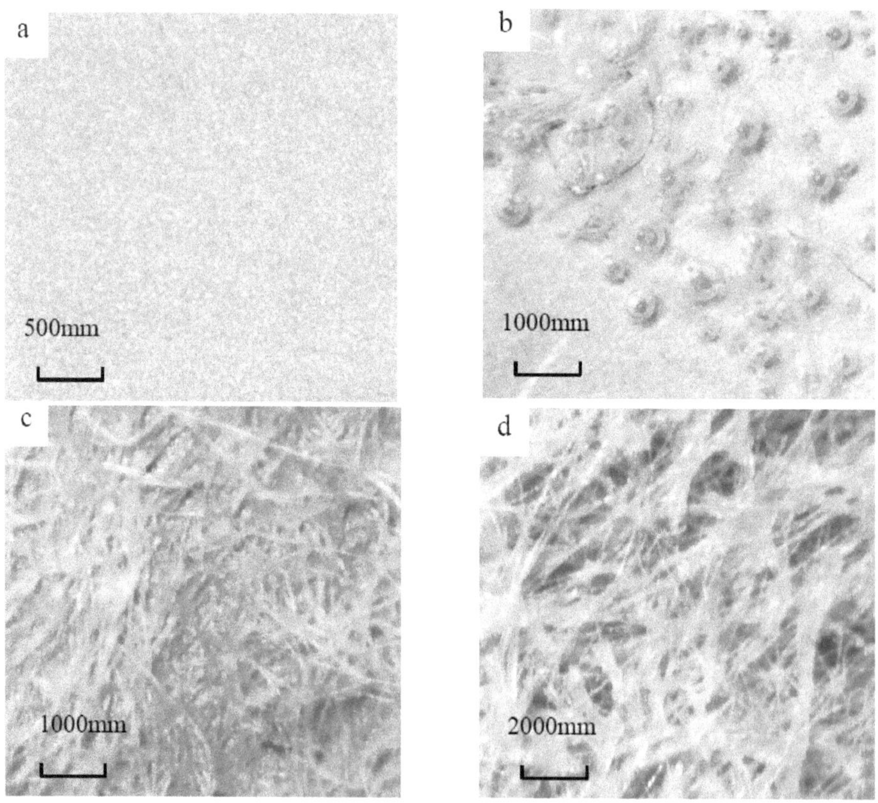

Figure 6: Topography of sample surface with catkins attachment over time, a: 0 h; b: 0.5 h; c: 1 h; d: 2 h

Fig. 7 shows the existing form when there are oils on the surface of the sample. As can be seen from Fig. 7b, dirt is wrapped by oil, only heavier dirt could deposit beneath the oil layer to be in contact with the test piece in order to complete this phase of growth, verifying the above analysis about that dirt gains weight slower than others with oil on the surface.

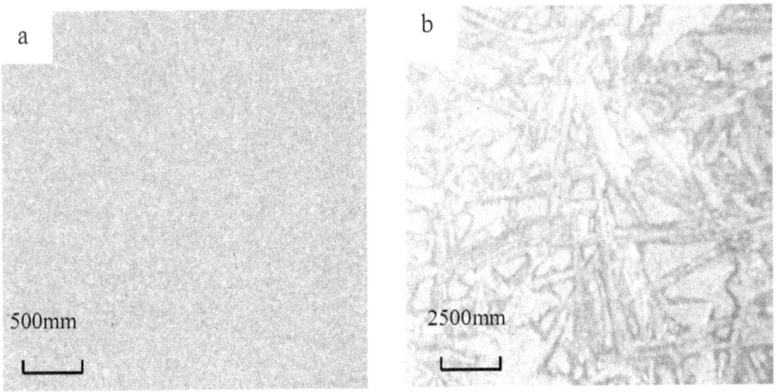

Figure 7: Topography of sample surface with oil attachment over time, a: 0 h; b: 4 h

Fig. 8 illustrates the change with time of sample surface covered with cleaning agent. Dirt is wrapped by oily cleaning agent as shown in Fig. 8b, which demonstrates that the inhibition of dirt by cleaning agent is also because of the cut of contact between the thin film and the test pieces. At the same time, dirt precipitated from the cleaning agent is relatively small compared to other dirt, indicating that the cleaning agent has a potent inhibitory effect on dirt growth.

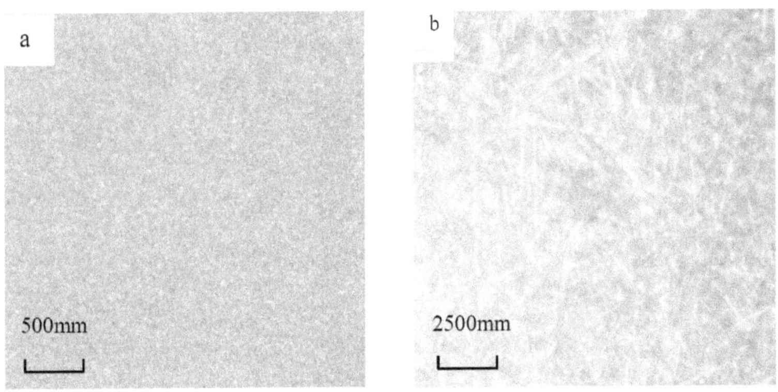

Figure 8: Topography of sample surface with cleaning agents over time, a: 0 h; b: 4 h

4 Conclusions

The source of aircraft air-conditioning heat exchanger dirt are mainly air pollutants and soil particles, including ions such as Na^+, NH_4^+, K^+, Mg^{2+}, Ca^{2+}, Cl^-, NO_2^-, Br^-, NO_3^- and SO_4^{2-}.

During the evaporation process, dirt nuclei generated randomly in the thin film. But when the test piece surface has catkins or other large particulates, it will augment adsorption sites of nucleation. Thereby the growth of dirt was significantly increased, the density of the dirt as well.

When there are large particles like catkins on the surface of specimens, the crystallization dirt grows the fastest, while when the specimens are covered with oil or residual cleaning agents, fouling growth has been somewhat suppressed for the reason that residual cleaning agents and oil prevent test piece from the atmosphere, it's difficult for the dirt nucleation to deposit on the surface of the sample.

Acknowledgment: This study was supported by the Major Science & Technology Research Project of Civil Aviation of China (No. MHRD20140110).

References

[1] Chen M.Y., Zhang X.Y., Cai W., Operating Analysis of Airplane Air Conditioning Heat Exchanger[J]. J. Shanghai Univ. Eng. Sci., 2008, 04:310-314

[2] Li L.K., Shooting of "B737-300/500 A/C System Cannot Cooling"[J], Aviat. Maint. Eng., 2004, 3:48-49

[3] Yiantsios S.G., Karabelas A.J., Deposition of micron-sized particles on flat surfaces: Effects of hydrodynamic and physicochemical conditions on particle attachment efficiency[J], Chem. Eng. Sci., 2003, 58(14):3105-3113

[4] Xu Z.M., Zhang Z.B., Guo W.Z., et al., A theoretical model of composite fouling of particulate and crystallization[J], J. Eng. Thermophys., 2006, 27(z2):81-84

[5] Zhang Z.B., Xu Z.M., Qiu Z.B., Experimental investigation on effect of surface and flow characte-ristics on the induction period of fouling[J], J. Eng. Thermophys., 2009, 30(1):144-146

[6] Sheikholeslami R., Nucleation and kinetics of mixed salts in scaling[J], Aiche J., 2003, 49(49):194-202

[7] Kim W.T., Bai C., Cho Y.I.,A study of CaCO3 fouling with a microscopic imaging technique[J], Int. J. Heat Mass, 2002, 45(3):597-607

[8] Yang F.,Tan J.,Zhao Q., et al., Characteristics of $PM_{2.5}$ speciation in representative megacities and across China[J], Atmos. Chem. Phys., 2011, 11: 5207-5219

[9] Ravishankara A.R., Heterogeneous and Multiphase Chemistry in the Troposphere[J], Science, 1997, 276(5315):1058-1065

[10] Saxena P., Hildemann L.M., Water-soluble organics in atmospheric particles: A critical review of the literature and application of thermodynamics to identify candidate compounds[J], J. Atmos. Chem., 1996,24(1):57-109

[11] Limbeck A., Puxbaum HOr., ganic acids in continental background aerosols[J], Atmos. Environ., 1999, 33: 1847-1852

[12] Shi Y.Y., The electrochemical studies of atmospheric corrosion of typical metals[D]., Zhejiang University, 2008
[13] Liu Y.D. Influence of Behavior of Metallic Heat Transfer Surface on the Formation Characteristics of Crystalline Fouling[D]., Shandong University, 2008
[14] Bansal B., Chen X.D., Müller-Steinhagen H., Analysis of 'classical' deposition rate law for crystallization fouling[J], Chem. Eng. Process., 2008, 47:1201-1210
[15] Zhou X., Cai X.J., Liu X.C., et al., Analysis of heat exchanger fouling formation mechanism and influencing factors[J], Petro Chem. Equip., 2014, 43(1):84-88

Huaxiang He*, Xiaoya Deng, Xiangnan Zhou, Zhenzhen Ma, Xinmin Xie

The Effect of Meteorological Factors on the Water-pollution Accident Risk Prediction of River-road Intersection by Using Binary Logistic Regression Model

Abstract: Based on the Binary Logistic regression model, this paper proposed a method to identify the probability of the sudden water pollution accident caused by hazardous materials transportation accident. This paper located 43 river-road intersection systems by using ArcGIS spatial database in Songhua River Basin. This paper considered the meteorological factors as input variable for the establishment of Binary Logistic regression model to analyze the accident probability and used statistics of accidents to check this model. This paper selected the daily data of temperature, precipitation, wind speed and relative humidity from 48 national meteorological stations and traffic accidents of 3 provinces from 2006 to 2008 to simulate the probability of accidents in 43 river-road intersections. The results show that low risk accuracy of the sudden water pollution accident probability was 53.8% and high risk accuracy was 70.5%, with the total reaching 65.9%. This paper plotted an accident risk warning map of the river-road intersection system of Songhua River basin which could provide technical support for water environment management departments in their efforts to shift from emergency response management to proactive risk management in dealing with water pollution accidents.

Keywords: binary logistic regression model, water-pollution accident probability, river-road intersection system, meteorological factors.

1 Introduction

With the rapid economic development, the water pollution problem was increasingly serious. According to the statistics from 2000 to 2011 [1-6], there were 1176 accidents which caused sudden water pollution across the whole country, among which,

***Corresponding author: Huaxiang He,** State Key Laboratory of Hydroscience and Engineering, Department of Hydraulic Engineering, Tsinghua University, China Institute of Water Resources and Hydropower Research, Beijing, China, E-mail: hehx@iwhr.com
Xiaoya Deng, Zhenzhen Ma, Xinmin Xie, State Key Laboratory of Simulation and Regulation of Water Cycle in River Basin, China Institute of Water Resources and Hydropower Research, Beijing, China
Xiangnan Zhou, Yellow River Engineering Consulting Co.,Ltd., Zhengzhou, China

hazardous materials transportation (HMT for short) accidents caused the most harmful sudden water pollution compared with others, which included accidents caused by emission of industrial wastewater beyond the permissible limit values or without treatment as well as leakage accidents in pipeline. Recently, the HMT accident had a tendency to increase year by year, such as Han River pollution accident in 2011, Qidao urban river pollution accident in 2012, Zi River pollution accident in 2013 and so on. Risk assessment is the fundamental work for risk decision-making and management. Due to the uncertainty of risk assessment, in which statistical approaches were commonly used [7-12].

The accidents posed great potential threat for social public security due to the flowing feature of water. The study was relatively weak in accident prediction of river-road intersection (shorted as RRI) system which is defined as bridges or roads across rivers or along rivers. Based on statistical principles, researchers did some analysis for traffic accident on frequencies, casualties, economic losses and traffic flow [13-16]. The studies about sudden water pollution caused by HMT accidents were mainly based on statistical principles and analyzed the intrinsic mechanism of accidents. As suggested by Mannering et al. [17-18], random parameter approaches can analyze specified groups of observations. Those taking into account various meteorological factors were also more [19-24], especially precipitation and wind speed [25-29]. While, research considering temperature and relative humidity were actually rather rare. In addition, research applying to water environment management of the river basin was much rarer.

Taking the Songhua River basin as a main study area, which has intensive river network crossing 10 national roads, this study includes an analysis of the relationship between the meteorological factors and the accident rate followed by a method of the accident risk prediction in RRI system by Binary Logistic regression Model. The accident risk prediction can provide technical support for water environment management departments in their efforts to shift from emergency response management to proactive risk management in dealing with water pollution accidents [30].

2 Method

Binary Logistic regression is widely used in the risk prediction of traffic accidents [16,31-33]. It can describe quantitatively the results of dependent variables which are the probabilities of risks given by the values of the independent variables about two scenarios. In this study, the independent variables include several meteorological factors, such as temperature, precipitation, wind speed, relative humidity. The dependent variable is the probability of the sudden water pollution accident caused by an accident of dangerous goods transportation.

The formula of probability is as follows:

$$P=\frac{\exp(\beta_0+\beta_1 X_1+\beta_2 X_2+\cdots+\beta_m X_m)}{1+\exp(\beta_0+\beta_1 X_1+\beta_2 X_2+\cdots+\beta_m X_m)} \quad (1)$$

In which, P is the dependent variable defined as the probability of the sudden water pollution accident, $P=P(Y=1|X_1,X_2,\cdots,X_m)$ stands for high risk and $P=P(Y=0|X_1,X_2,\cdots,X_m)$ stands for low risk. $X_1,X_2,\cdots,X_m$ are the independent variables. β_0 is regression intercept. When $X=0$, $\beta_0=\ln(X_{highrisk}/X_{lowrisk})$, $\beta_1,\beta_2,\cdots,\beta_m$ are partial regression coefficient. When X is constant, $\log itP$ changes along with the independent variable each unit.

The logistic regression model can be expressed as follows:

$$\log itP=\ln\frac{P}{1-P}=\beta_0+\beta_1 X_1+\beta_2 X_2+\cdots+\beta_m X_m \quad (2)$$

Firstly, the probability treshold value (abbreviation - Tv) should be determined by using Logistic regression model [34]. When P is greater than or equal to Tv, the risk is high. On the contrary, the risk is low.

3 Case Study

Songhua River is one of the seven major rivers in China, which has 561.2 thousand km^2. Songhua River has two sources, one is from north to south called Nen River and the other is from south to north called the Second Songhua River. After converging, the river flows from southwest to northeast into Heilongjiang River. Songhua River flows through the following administrative areas such as Heilongjiang Province, Jilin Province and Neimenggu Autonomous Region (provinces of Hei, Ji and Nei for short). The Songhua River basin is one of the most important places in Northeast Area Revitalization Plan. In 2005 and 2006, there were two major water pollution accidents drawing extensive attention of the international community.

3.1 Data Collection

1. The meteorological factors were downloaded from the China meteorological data sharing service network (http://cdc.cma.gov.cn/home.do), which provides meteorological factors from 48 national weather stations such as temperature, precipitation, wind speed and relative humidity. The Series of meteorological factors are daily data from 2006 to 2009.
2. This paper collected traffic map at http://www.onegreen.net with scale of 1:7950 thousand shapefiles of ArcGIS about water function area and water system images from Songliao River Conservancy Commission of China. Songhua River basin contains 10 main national highways and 43 river-road intersections, as shown in Fig. 1 and Table 1.

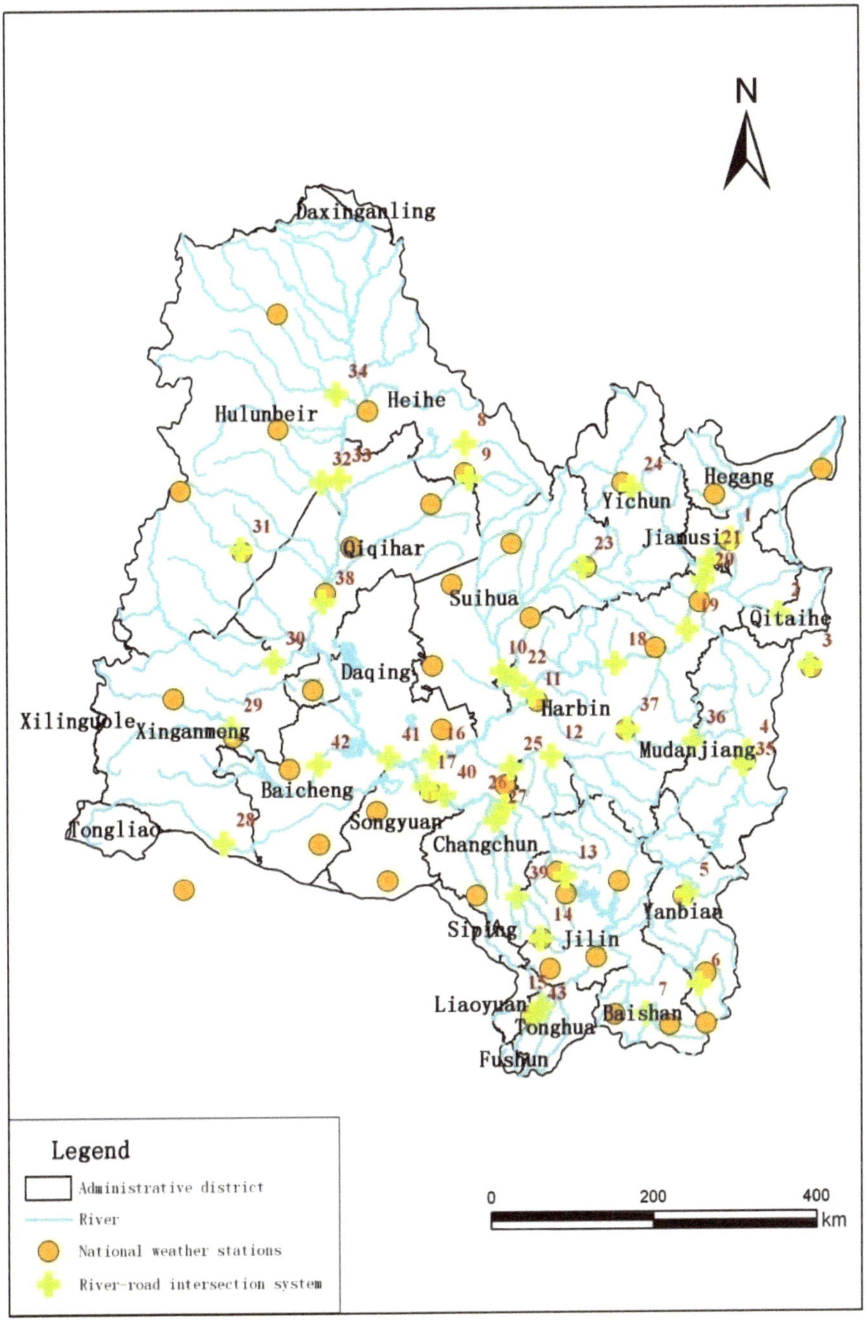

Figure 1: RRI system in Songhua River basin

Table 1: Basic Information of RRI System

Serial number	Road Code	Road name	The number of intersections	River name	River long/km	Intersection code
1	G201	XAQ-JMS	1	The main Songhua River	56	G201-1
2		HN-QTH	1	Woken River	57	G201-2
3		LK-MDJ	1	Mudan River	119	G201-3
4		NA-DH	1	Mudan River	201	G201-4
5		DH-FS	2	Mudan River	276	G201-5
6				Erdaobai River		G201-6
7		FS-JY	1	Toudao River	129	G201-7
8	G202	SW-BA	1	Nemoer River	185	G202-1
9		BA-KD	1	Wuyuer River	45	G202-2
10		LX-HEB	1	Hulan River	79	G202-3
11		HEB-YS	2	Songhua River	129	G202-4
12				Lalin River		G202-5
13		YS-JLS	1	The second Songhua River	127	G202-6
14		YJ-PS	1	Yinma River	116	G202-7
15		PS-MHK	1	Huifa River	65	G202-8
16	G203	ZY-SY	1	The main Songhua River	53	G203-1
17		SY-QGELS	1	The second Songhua River	8	G203-2
18	G221	BX-FZ	1	The main Songhua River	120	G221-1
19		FZ-YL	1	The main Songhua River	103	G221-2
20		YL-JMS	2	The main Songhua River	105	G221-3
21				Tangwang River		G221-4
22	G222	HEB-HL	1	Hulan River	35	G222-1
23		QA-TL	1	Hulan River	48	G222-2
24		TL-YC	1	Tangwang River	109	G222-3
25	G102	CC-SC	3	Lalin River	206	G102-1
26				The second Songhua River		G102-2
27				Yinma River		G102-3
28	G111	TL-KYZQ	1	Huolin River	240	G111-1
29		TQ-WLHT	1	Taoer River	114	G111-2
30		WLHT-ZLTQ	1	Chuoer River	115	G111-3
31		ZLTQ-ZLT	1	Yalu River	192	G111-4
32		ARQ-MLDW	1	Nuomin River	143	G111-5
33		MLDW-JGDQ	2	Nen River	288	G111-6
34				Gan River		G111-7
35	G301	SFH-MDJ	1	Mudan River	183	G301-1
36		MDJ-LH	1	Mayi River	24	G301-2
37		LH-SZ	1	Mayi River	163	G301-3
38		QQHEMLS-GN	1	Nen River	27	G301-4
39	G302	JL-CC	1	Yinma River	119	G302-1
40		NA-SY	1	The second Songhua River	91	G302-2
41		SY-DA	1	Nen River	66	G302-3
42		DA-BC	1	Taoer River	135	G302-4
43	G303	LH-MHK	1	Huifa River	30	G303-1

3. The traffic statistics were downloaded from traffic sharing network (http://218.247.138.157: 8080/index1.jsp), which provides some traffic statistics from 2006 to 2009 road mileage and traffic flow of provinces of Hei, Ji and Nei from the report of the Second National Highway Survey. The accuracy of traffic flow was relatively low because the national highway traffic flow detection equipments were in lower density layout and historical traffic flow data was incomplete. The hypothesis of this study was that highway traffic flow was constant in the year.
4. The number of the sudden water pollution accidents caused by HMT accidents was collected from China's transportation yearbook, China's water resources bulletin and survey statistics.

High risk is defined as 3 (or more) accidents per day. The validation period was from 2006 to 2008 and the prediction period was from 2009 to 2010 in Binary Logistic regression.

3.2 Data Analysis

1. Temperature. The mean temperature had a descending trend from 2006 to 2010. The average monthly temperature was the lowest in January with -12 ~ -17°C and the highest in June with 19~25°C. The average annual temperature of 2007 was higher than in the other years. The temperature difference can exceed 58°C in whole year of 2010 (See Fig. 2).
2. Precipitation. The precipitation had a descending trend from 2006 to 2010 and was uneven distributed the whole year round. It mainly concentrated from June to September. The precipitation of spring flood season in 2006 was lowerthan in the other years. The proportions of precipitation in winter were 2% in 2006 and 7% in 2007. There was no snowfall in January and February of 2009 (See Fig. 3).
3. Wind speed. During 2006-2010, the wind speed was not significant in whole year (See Fig. 4). The average of daily wind speed was 2.0~2.3 m/s, the maximum was 2.9 m/s and the minimum was 1.4 m/s.
4. Relative humidity. Relative humidity maintained among 36%-83% from 2006 to 2010, with an average of 60%. It was higher in summer and winter, so was the precipitation. The minimum value of relative humidity occurred in December 2010. See Fig. 5.
5. The number of RRI system accidents had a decreasing trend from 2006 to 2010. Smaller number of accidents occurred in February and October and bigger in December. On fridays, saturdays, sundays was generally more accidents than on the other dates.Especially, the number of accidents was higher in May 2006 tahn during the other periods. The accidents decreased by 21% in winter 2008.

Thus, the number of accidents was negatively correlated to the temperature. The effect of the wind speed and relative humidity on the number of accidents can be offset by each other when there were small fluctuations in temperature. The low temperature (below zero degrees Celsius), heavy precipitation (deeper than 15 mm), high relative humidity (above 50%) and strong wind (wind speed greater than 1.0 m/s) were closely related to the number of accidents. The influence on the number of accidents is expressed by the inequality as follows: precipitation > relative humidity > wind speed > temperature.

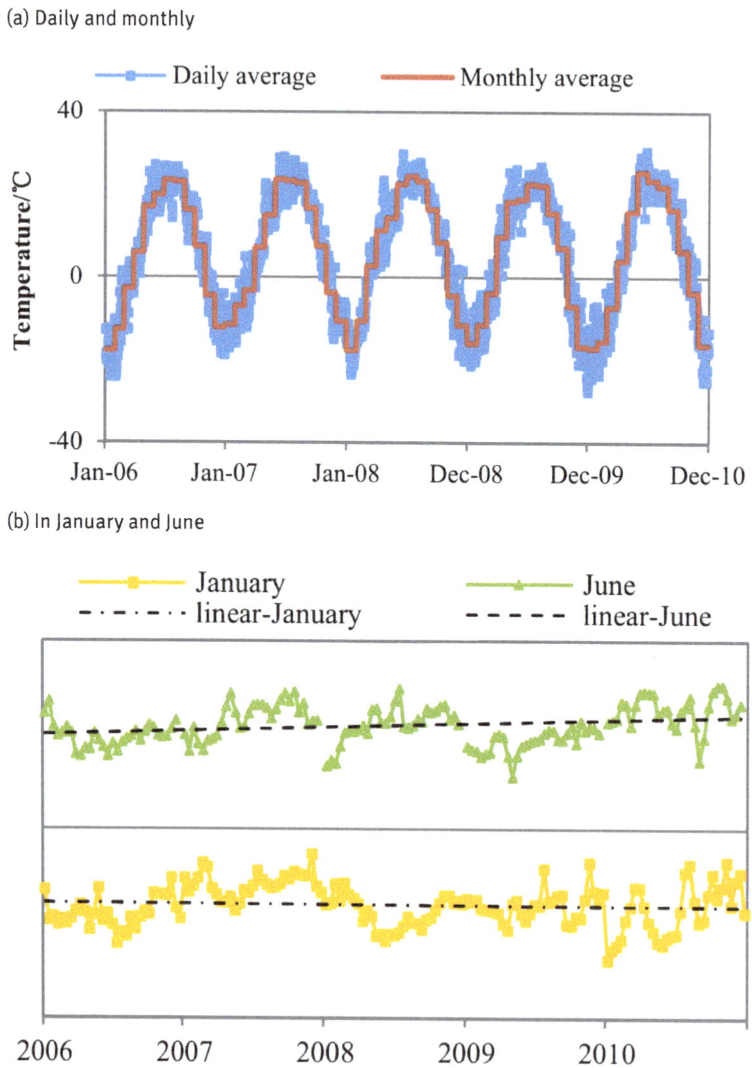

Figure 2: Temperature of HEB station from 2006 to 2010

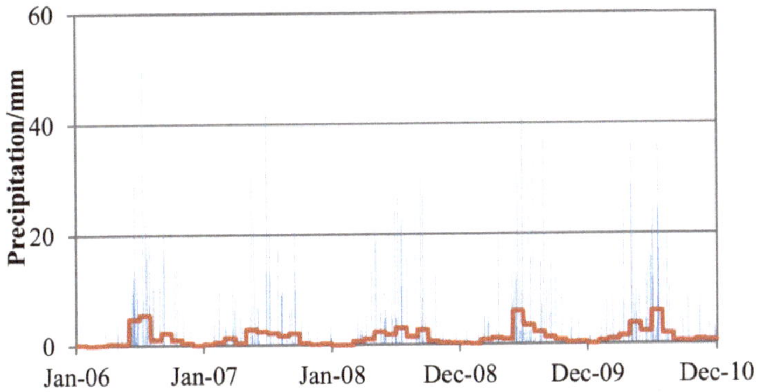

Figure 3: Precipitation of HEB station from 2006 to 2010

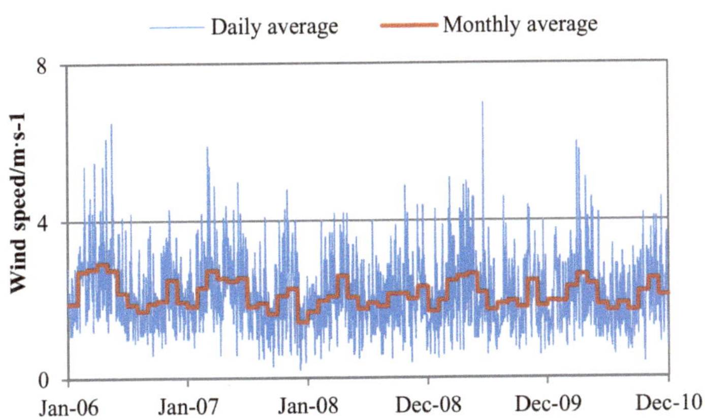

Figure 4: Wind speed of HEB station from 2006 to 2010

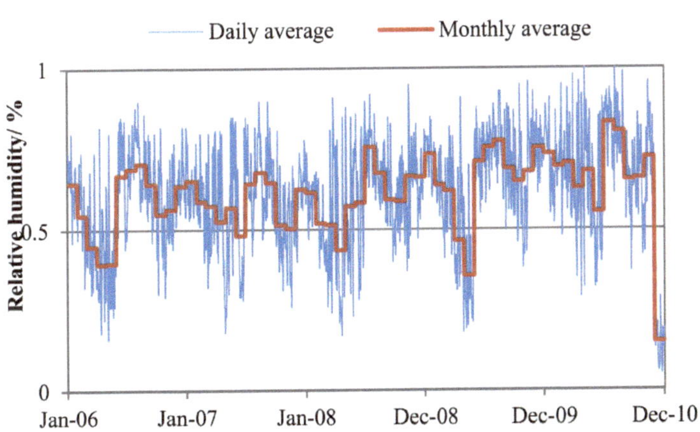

Figure 5: Wind speed of HEB station from 2006 to 2010

3.3 Model Establishment

Binary Logistic Regression model was established to simulate risk of RRI system from DTC accident by SPSS version 11.0 [35]. Steps are as follows:

1. Independent variables selection. This paper used the Spearman correlation to calculate the correlation coefficient between meteorological factors and made some selections to weed out highly correlated variables.
2. Standardization of independent variables is as follows:

$$f(n\Delta t)=\frac{x-\overline{x}}{\sigma}. \tag{3}$$

$$\sigma^2=\frac{n\sum x^2-(\sum x)^2}{n(n-1)}. \tag{4}$$

In which n is the number of samples, x is independent variables, $\overline{x}$ is the average of independent variables, σ is mean square error of independent variables.

3. Binary Logistic regression model establishment. If all the independent variables have 'Enter' into the Binary Logistic regression model, it cannot exclude anything in the level of significance less or equal to 0.05. If all the independent variables are 'Forward LR' or 'Backward LR' into the Binary Logistic regression model, it can exclude something in the level of significance less or equal to 0.05.
4. ROC curve to determine the classification standard. True Positive Rate means sensitivity as the ordinate in X-Y coordinates. False Positive Rate means probability as the abscissa in X-Y coordinates. Usually the TPR is larger, the model precision is higher.
5. Results comparison. The model parameters are determined by comparison of three methods. 'Enter' method can force all the variables into the model in case of fewer cases. 'Forward' method and 'Backward' method can eliminate variables of common problems. The difference between 'Forward' method and 'Backward' is that variables are into or out one by one.

3.4 Model Application

1. **Independent variables were selected:** There were 4384 samples including temperature, precipitation, wind speed and relative humidity. Descriptive and correlation analyses were conducted. No significant correlation was found between wind speed and temperature inspection in 95% confidence intervals. Others were with lower correlation coefficients. See Table 2.

Table 2: Correlation coefficients between meteorological factors

The correlation coefficient	Temperature	Precipitation	Wind speed	Relative humidity
Temperature	1.000	0.153**	-0.030	0.079**
Precipitation	0.153**	1.000	0.155**	0.478**
Wind speed	-0.030	0.155**	1.000	-0.261**
Relative humidity	0.079**	0.478**	-0.261**	1.000

**At confidence intervals (double sides) of 0.01, the correlation coefficient is significant.

2. **Model accuracy was determined:** When the classification standard was equal to 0.5, the model precision was highest. See Fig. 6. The threshold value was determined to be 0.5 in this study. When P was greater than or equal to 0.5, it indicates high risk of the sudden water pollution accident probability. Otherwise, it was low risk of the sudden water pollution accident probability. Low risk accuracy of the sudden water pollution accident probability was 53.8% and high risk accuracy was 70.5%, with the total reaching 65.9%.

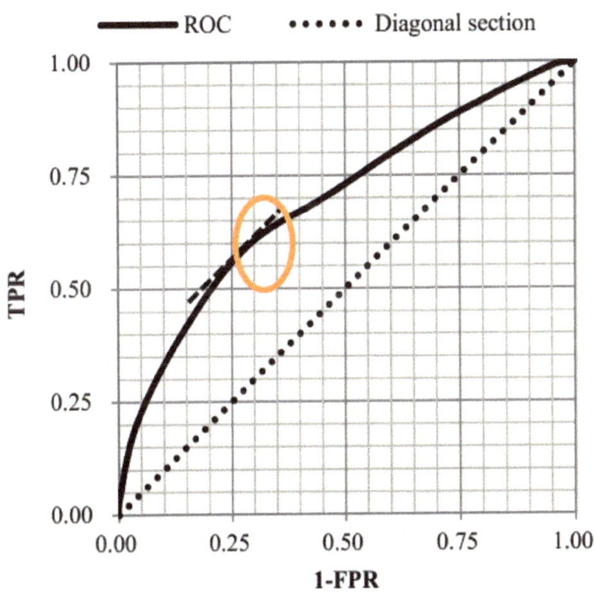

Figure 6: ROC curve of Logistic regression model

G201 - LX - HEB as an example, the prediction model formula was as follows:

$$P = \frac{\exp(0.254 + 1.237 \times T + 5.032 \times PE + 0.638 \times WS + 2.124 \times RH)}{1 + \exp(0.254 + 1.237 \times T + 5.032 \times PE + 0.638 \times WS + 2.124 \times RH)}. \tag{5}$$

In which, *T* is temperature, *PE* is precipitation, *WS* is wind speed, *RH* is relative humidity.

The accuracy was lower than 40% in April, May, July and November and lower than 20% in August 2008. Monthly samples statistics showed that there were 70% samples with the accuracy higher than 60%. See Fig. 7. Thus, the model proved to be better in predicting the risk probability of the sudden water pollution accident.

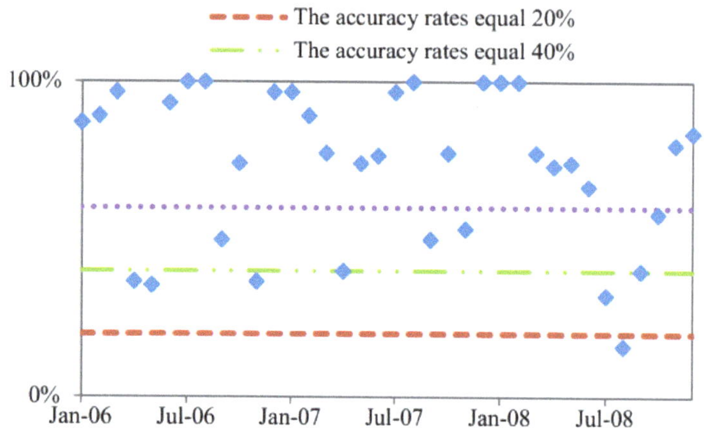

Figure 7: Distribution of the prediction accuracy by Logistic regression model

4 Conclusions

Harbin station was an example of the prediction results of February 2010 and April 2010, in which precipitation was higher than the average level by a factor of 4.5 times and 1.5 times respectively, should be low risk by prediction but the actual results of those were high risks. The prediction results for September, in which precipitation was lower than average level of 45%, should be high risk but the actual results of that was low risk. There were some prediction results shown in Fig. 8, which can express the risk condition of the spring flood season, flood season, normal season and icebound season respectively.

In April 2009, the proportion of high risk at G202-1, G201-7 was 19% based on RRI system, posing a threat to the water quality safety of the Rivers of Nemoer and Toudao, both were drinking water sources. At the same time, G111-3, G102-1, G102-2, G102-3 located at provincial boundaries between Hei and Ji provinces should be paid more attention, while controversies exist in terms of the responsibility division of water quality management.

In July 2009, the proportion of high risk had risen to 95% of RRI system. The safety inspection and supervision work should be carried out on RRI system in flood season. Meteorological and hydrological forecasting should provide data supports for making decision.

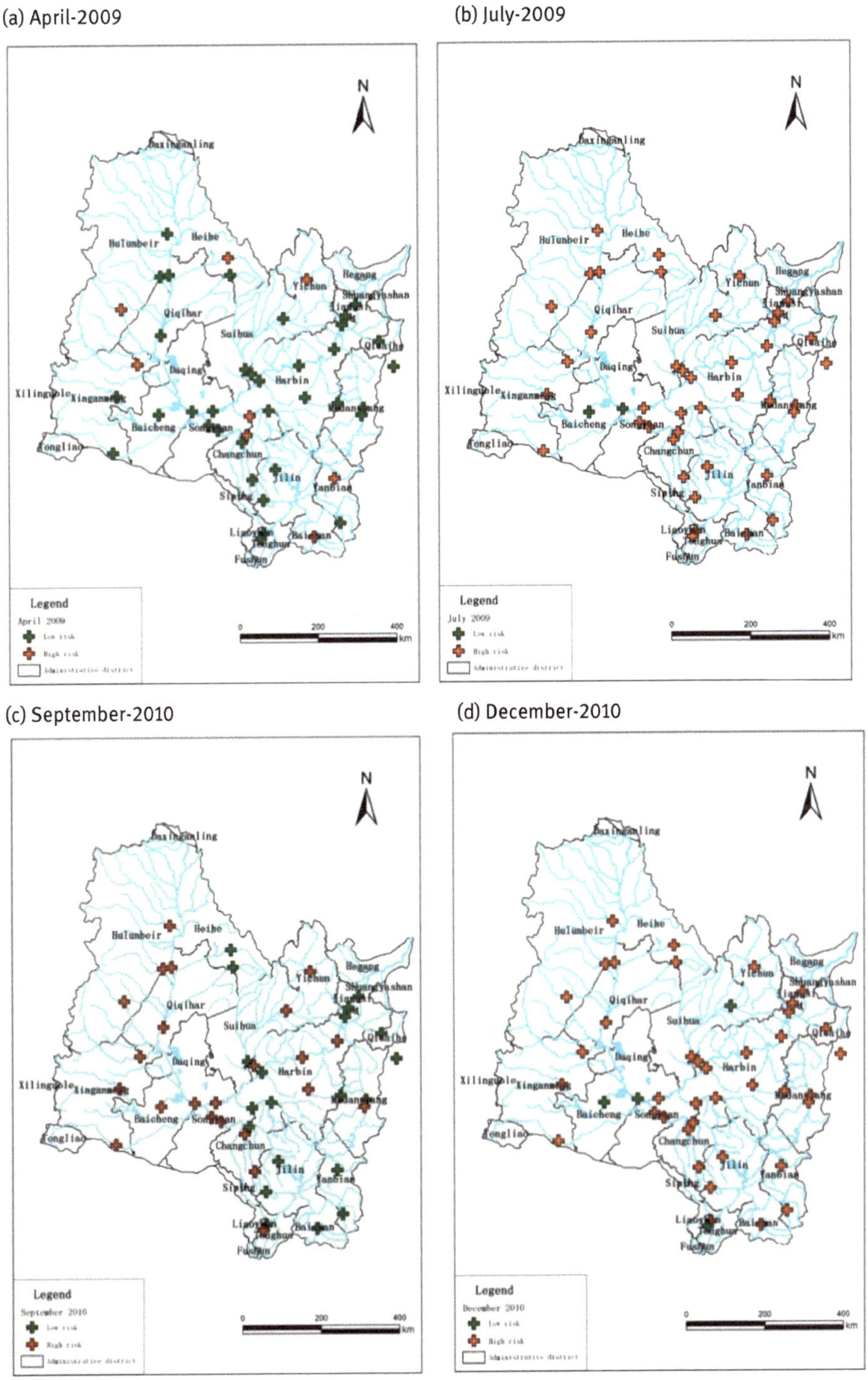

Figure 8: Risk early warning figure of river-road intersection in Songhua River basin

In September 2010, the proportion of high risk was 53% of river road intersection system where located at G111, G203, G222, G302.

In December 2010, the proportion of high risk was 91% of RRI system where located at G201-6, G201-7, G202-8, G303-1. Especially, Toudao River, Erdaobai River and Huifa River without frozen, when the weather was getting worse, the prevention level should be improved and the inspection of the area reach was strengthened.

To summarize accidents in RRI system were random, the Logistic regression model could predict risk probability quantitatively with meteorological factors. High risk occurred in flood season. After the flood season, the accident risk probability decreased significantly, but increased in icebound season caused by lower temperature, higher wind speed, ice cover and snowfall. Moreover, Songhua River basin distinguished from other river basins in China, by higher risk probability in spring flood season.

5 Discussion

The ministry of water resources of China or the ministry of environmental protection of China performs their own functions respectively: the former does not 'go ashore' and the latter does not 'go underwater'. Once the sudden water pollution in HMT accidents goes into the water body, emergency relief is required from the ministry of water resources. Therefore, this study can provide emergency supervision in the decision support technology for the Ministry of Water Resources.

In this paper, the Binary Logistic regression model is used to calculate the occurrence probability of sudden water pollution accident quantitatively with temperature, precipitation, wind speed, relative humidity as the independent variables.

Taking the RRI system of the Songhua River basin as a case study, the definition of the risk threshold value is proved that the research can be referenced in some periods. In addition, runoff was perhaps another direct factor which influences the probability of sudden water pollution accident. Due to the lack of direct observation in runoff, hydrological model or other tools should be applied in simulating runoff. We will analyze and demonstrate this aspect in the follow-up studies.

Acknowledgment: This study received financial support from State water pollution control and management of major scientific and technological special issue (2012ZX07201-006; 2008ZX07207-006) and Special Funds for Public Industry Research Projects of the National Ministry of Water Resources (201501013).

References

[1] Ai H.Y., Liu T.W..Statistical review of the major unexpected water contamination incidents at home in the period from 2000 to 2011. Journal of Safety and Environment (In Chinese) [J], 2013, 13(4): 284-288.

[2] Li J., Shao S.G., QIANG R.R..Technical about collection and analysis of hazardous chemicals transportation accidents for emergency in water environment sensitive sections. Highway Traffic Science and Technology: Application Technology Edition (In Chinese) [J], 2010, 10: 480-482.

[3] An Y., Li S.C.. Statistics of environmental events in China during period from May to June in 2011. Journal of Safety and Environment (In Chinese) [J], 2011, 11(4): 264-268.

[4] An Y., Li S.C.. Statistics of environmental events in China during period from May to June in 2012. Journal of Safety and Environment (In Chinese) [J], 2012, 12(4): 263-268.

[5] An Y., Li S.C.. Statistics of environmental events in China during period from November to December in 2012. Journal of Safety and Environment (In Chinese) [J], 2012, 13(1): 280-284.

[6] An Y., Li S.C.. Statistics of environmental events in China during period from May to June in 2013. Journal of Safety and Environment (In Chinese) [J], 2012, 13(2): 280-284.

[7] Uricchio V.F., Giordano R., Lopez N.. A fuzzy knowledge-based decision support system for groundwater pollution risk evaluation. Journal of Environmental Management[J], 2004, 73: 189-197.

[8] Kentel E., Aral M.M.. Probabilistic-fuzzy health risk modeling. Stochastic Environmental Research and Risk Assessment[J], 2004, 18: 324-338.

[9] Kentel E., Aral M.M.. 2D Monte Carlo versus 2D fuzzy Monte Carlo health risk assessment. Stochastic Environmental Research and Risk Assessment[J], 2005,19: 86-96.

[10] Wang X.L., Zhang J.. A nonlinear model for assessing multiple probabilistic risks: a case study in South five-island of Changdao National Nature Reserve in China. Journal of Environmental Management[J], 2007, 85 (4): 1101-1108.

[11] Matthies M., Giupponi C., Ostendorf B.. Environmental decision support systems: current issues, methods and tools. Environmental Modelling & Software[J], 2007, 22: 123-127.

[12] Volk M., Lautenbach S., Delden H., Lachlan T., Newham H., Seppelt R.. How can we make progress with Decision Support Systems in Landscape and River Basin Management? Lessons learnt from a comparative analysis of four different decision support systems. Environmental Management[J], 2010, (46): 834-849.

[13] Luo H., et al. The Relationship Between Road Traffic Crashes and Meteorological Condition with Construction of Its Road Weather Warning Model. Journal of applied meteorological science (In Chinese) [J], 2007, 18(3): 350-356.

[14] Liu Y.X., et al. Study on the relationship between bad weather and traffic accident in Heilongjiang Province. Communications Science and Technology of Heilongjiang (In Chinese) [J], 2010, 10: 132-135.

[15] Xu C.C., et al. Real time crash risk prediction model on freeways under nasty weather conditions. Journal of Jilin University(Engineering and technology edition) [J], 2013, 43(1): 68-73.

[16] Cerwick D.M., et al. A comparison of the mixed logit and latent class methods for crash severity analysis. Analytic Methods in Accident Research[J], 2014, 3: 11-27.

[17] Mannering F. L., Bhat C. R.. Methodological frontier and future directions. Analytic Methods in Accident Research[J], 2014, 1: 1-22.

[18] Mannering F. L., Shankar V., Bhat C. R.. Unobserved heterogeneity and the statistical analysis of highway accident data. Analytic Methods in Accident Research[J], 2016, 11: 1-16.

[19] Smith K.. How seasonal and weather conditions influence road accidents in Glasgow. Scott. Geog. Mag. [J], 1982, 98(2): 103-114.
[20] Edwards J.B.. The relationship between road accident severity and recorded weather. Journal of Safety Research[J], 1998, 29(4): 249-262.
[21] Andreescu M. P., Frost D.B.. Weather and traffic accidents in Montreal, Canada. Clim. Res[J], 1998, 9: 225-230.
[22] Andrey J., Mills B, Leahy, M., Suggett, J.. Weather as a chronic hazard for road transportation in Canadian cities. Natural Hazards[J], 2003, 28: 319-343.
[23] Golob T. F., Recker W.W.. Relationships among urban freeway accidents, traffic flow, weather, and lighting conditions. Journal of Transportation Engineering[J], 2013, 129(4): 342-353.
[24] Md. Shohel Reza Amin, Alireza Zareie, Luis E. Amador-Jiménez. Climate change modeling and the weather-related road accidents in Canada. Transportation Research Part D: Transport and Environment[J], 2014, 32: 171-183.
[25] Sherretz L.A., Farhar B.C.. An analysis of the relationship between rainfall and the occurrence of traffic accidents. J. Appl. Meteorol[J], 1978, 17: 711-715.
[26] Andrey J., Sam Y.. A temporal analysis of rain-related crash risk. Accident Analysis & Prevention[J], 1993, 25(4): 465-472.
[27] Eisenberg D.. The mixed effects of precipitation on traffic crashes. Accident Analysis & Prevention[J], 2004, 36(4): 637-647.
[28] Young R.K., Liesman J., 2007. Estimating the relationship between measured wind speed and overturning truck crashes using a binary logit model. Accident Analysis & Prevention[J], 2013, 39(3): 574-580.
[29] Loukas D., Elenil.V.. Fuzzy modeling of freeway accident duration with rainfall and traffic flow interactions. Analytic Methods in Accident Research[J],2015, 5(6): 59-71.
[30] Meng W.. Construction of technology system for watershed water pollution treatment and management: exploration and practice of the major water program in Liaohe River basin. Engineering Sciences in China[J], 2013, 15(3): 4-10.
[31] Anastasopoulos P. C., Mannering F.L.. An empirical assessment of fixed and random parameter logit models using crash- and non-crash-specific injury data. Accident Analysis & Prevention[J], 2011, 43: 1140-1147.
[32] Shaheed M.B., Gkritza K., et al. A mixed logit analysis of two-vehicle crash severities involving a motorcycle. Accident Analysis & Prevention[J], 2013, 61: 119-128.
[33] Abegaz T., Berhane Y., et al. Effects of excessive speeding and falling asleep while driving on crash injury severity in Ethiopia: A generalized ordered logit model analysis[J]. Accident Analysis & Prevention, 2014, 71: 15-21.
[34] Jung S., Qin X., Noyce D.A.. Rainfall effect on single-vehicle crash severity using polychotomous response models. Accident Analysis & Prevention[J],2010, 42(1): 213-214.
[35] Zhang W.T. Statistical analysis tutorial by SPSS11[M].Beijing: Beijing Hope Electronic Press, 2002: 213-256.

Gabrielle E. Sherrer, Nirajan Dhakal*

Climate Change and Water Resources Management: An Integrated Assessment of Temporal Change in Population and Extreme Precipitation

Abstract: This research project involves the study of three counties within the Metro Atlanta Region in Georgia, USA with respect to two critical changes over the last six decades: extreme precipitation and population. The time-dependent changes in the extreme-rainfall return period were examined over the past 60-year span using the generalized extreme value (GEV) distribution for two 30-year blocks (1950-1979 and 1980-2009) of annual maximum daily precipitation. We found that across the state of Georgia, there was 16%, 22% and 28% increase in 25-year, 50-year and 100-year rainfall amounts respectively for the recent time period. All three counties have gone through population growth over the last 60 years - for one county the population has doubled, and for the other two counties the population has increased increased at least 5 times. Our findings have proven that as the population of an area increases, and we build upon natural resources, the use and necessity of water resource management tools (including the development of storm water infrastructures) increases with the urbanization of that area.

Keywords: extreme-rainfall, return period, generalized extreme value distribution, population growth, urbanization.

1 Introduction

Climate variability and change has the potential to cause significant impacts on our economic, ecological, social, and cultural resources. In a changing climate, civil infrastructures (such as dams, bridges, and culverts) are increasingly compromised during extreme precipitation events. However, much of our nation's infrastructure was developed assuming a non-varying climate [1,2]. As a result, we now see floods and infrastructure failure due to miscalculated design estimates. The increasing frequency and intensity of heavy rain events directly affect the efficiency and integrity of the stormwater infrastructure. For example, culverts (which carry water under the road) are engineered by studying the local hydrology (movement of water) and using intensity-duration-frequency curves, which calculate the likelihood of heavy rainfall

***Corresponding author: Nirajan Dhakal,** Environmental and Health Sciences Program, Spelman College, Atlanta, USA, E-mail: ndhakal@spelman.edu
Gabrielle E. Sherrer, Environmental and Health Sciences Program, Spelman College, Atlanta, USA

events of varying amounts and durations. This simply means that culverts can only accommodate a certain amount of water before they become incapacitated. When this happens, the road above floods, and in extreme cases the surrounding area can become unstable as soil around the culvert erodes over time. So if extreme rainfall events are becoming more frequent and more intense, culverts designed decades ago are becoming less efficient; and if growing populations are creating more road-stream crossings, some areas in the metro Atlanta area may be more vulnerable than others, depending on the changes in population and precipitation. In this paper we used a combined approach to view the relationship between change/increase in population and extreme precipitation risk in an effort to better define human-natural environment relationships vulnerable to increasing climate variability.

2 Data and Methodology

Our work involves the study of three counties within the Metro Atlanta Region; Fulton, DeKalb and Clayton, with respect to two critical changes over the last five decades: extreme precipitation and population. The 1950-2010 population for each county was compiled using census data. Total population per decade was estimated to identify pattern in population change. Daily precipitation data over the period of 1950–2009 from 21 stations across the state of Georgia was obtained from the Historical Climatology Network (HCN) database [3]. The dataset at each site was checked to make sure that they do not contain a large percentage of missing values, are short in length, e.g., just a few years, or, contain suspicious values in terms of quality.

Generalized Extreme Value (GEV) distribution [4] is usually used to statistically describe extreme rainfall events [5,6]. The GEV distribution is a flexible model that combines the Gumbel, Fréchet, and Weibull maximum extreme value distributions (three popular distributions used in the extreme value analysis). The GEV distribution has three parameters: location (μ), scale (σ) and shape (ξ). The cumulative distribution function of GEV is given by [5]:

$$F(x) = \exp\left\{-\left[1+\xi\left(\frac{x-\mu}{\sigma}\right)\right]^{-1/\xi}\right\} \tag{1}$$

We used GEV distribution to estimate the magnitude of extreme rainfall events using "fgev" routine in the "evd" library of the R statistical package as done by DeGaetano [6]. The routine uses the maximum likelihood (ML) method in fitting the data.

"Return period" (T) based on extreme events is commonly used to assess the service level (or the level of risk) of drainage systems by engineers and water resource managers (for example, settlements, urban areas, important infrastructure are designed for 50-100 years return-periods of rainfall or floods). Return period is an estimate of the likelihood of an event, such as a flood or a river discharge flow to occur. With the changing climate, the probability of exceeding the original design criteria based on return period also increases [7]. Hence, study of the impact of non-stationarity of the extreme precipitation events on the return-periods is crucial. For accessing the time-dependent changes in the return periods of the extreme-rainfall, the annual maximum series of the 30 extreme precipitation events from 1950-1979 and 1980-2009 were obtained separately. GEV distribution was fit to each series and the precipitation amounts corresponding to 10-, 25-, 50- and 100- years return period were estimated for all the 21 stations. The ratio of the rainfall amount (**r**) estimated based on the recent time period to the corresponding rainfall amount based on the old time period was evaluated to assess the temporal change in the rainfall risk:

$$r_i = (R_i)_{recent} / (R_i)_{old},$$

where, i = rainfall return period years (10, 25, 50, 100), Ri is the rainfall amount corresponding to the i-year return period

Bootstrap method [8] was used to assess the statistical significance of the change or uncertainty.

3 Results

Population change results are presented in Fig. 1. We can see that all the three counties have gone through population increase per each decade. Over the last 60 years (between 1950 and 2010) the population of Fulton county has almost doubled, population of DeKalb county has increased by 5 times, and population of Clayton county has increased by 11 times.

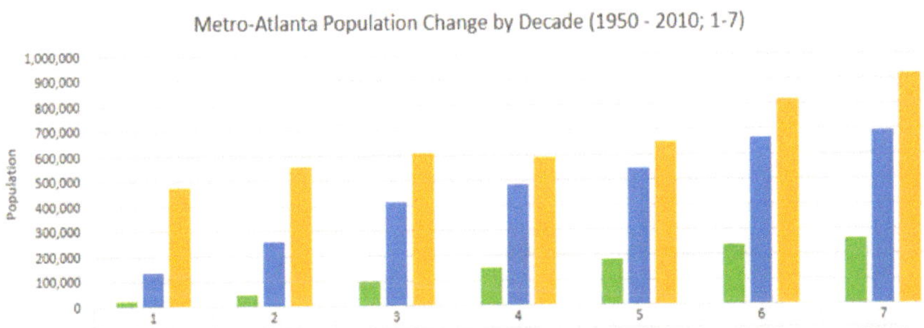

	1	2	3	4	5	6	7
Clayton	22,872	46,365	98,043	150,357	182,052	236,517	259,467
DeKalb	136,395	256,782	415,387	483,024	545,837	665,865	691,893
Fulton	473,572	556,326	607,592	589,904	648,951	816,006	920,581

Figure 1: Population per decade for each of the counties from 1950-2010; 1950-1959 is represented by 1, 1960- 1969 by 2 and so on.

To assess the relationship of change in population with the change in precipitation risk, r_{10}, r_{25}, r_{50} and r_{100} are obtained and presented in Table 1, and Fig. 2 and 3.

Table 1: Statistical summary of temporal change in extreme rainfall risk in GA

	10-year	25-year	50-year	100-year
Maximum	1.06	1.96	2.22	2.54
Minimum	0.88	0.78	0.70	0.62
Average	1.11	1.16	1.22	1.28
75% Quartile	1.18	1.30	1.35	1.41
Median	1.14	1.11	1.11	1.08
25% Quartile	0.96	0.99	1.02	1.03

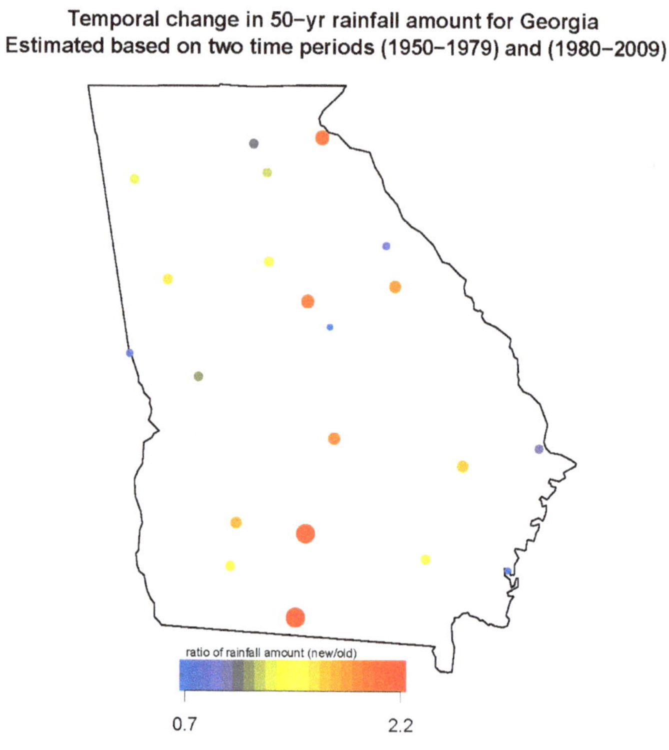

Figure 2: Ratio of 50-yr rainfall amount (new/old)

Statistical summary of the ratios (r) for 21 stations are listed in Table 1. The range of change across the stations is from 0. 6 to 2.5 times of the original value or the old value (Table 1). For 16 stations out of 21 (or for 76% of stations), an increase in both 50-yr and 100-yr rainfall amount was observed for the recent time period. For 8 stations, at least 25% increase in both 50-yr and 100-yr rainfall amount was observed for the recent time period. An increase of 50% or more was observed for 3 stations in the 50-yr and 4 stations in the100-yr rainfall amount. Although spatial pattern of the change is not discernible, however increase is more distinct in eastern and southern regions of the state (Fig. 2 and 3).

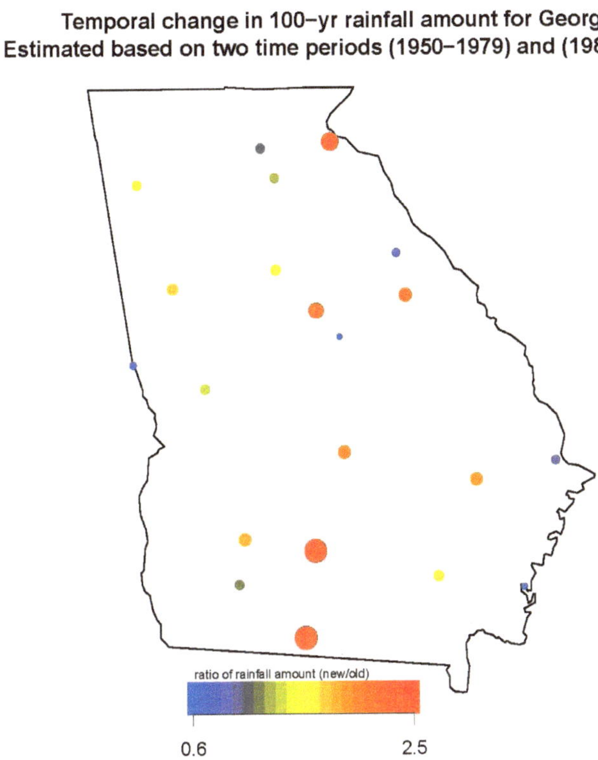

Figure 3: Ratio of 100-yr rainfall amount (new/old)

4 Discussion and Conclusions

This research project involves study of three counties within the Metro Atlanta Region in USA with respect to two critical changes over the last six decades: extreme precipitation and population. The time-dependent changes in the extreme-rainfall return period are examined over the past 60-year-span using the generalized extreme value (GEV) distribution for two 30 year blocks (1950-1979 and 1980-2009) of annual maximum daily precipitation. There has been in average 16%, 22% and 28% increase in 25-year, 50-year and 100-year rainfall amounts respectively for the recent time period. All three counties have gone through population growth over the last 60 years.

The purpose of this work was to illustrate the relationship between population change and the extreme precipitation risk. A large population easily suggests the presence of more roads, buildings, rooftops and other impervious land area. As sewage drains, pipes, culverts and bridges are implemented to manage storm water, the data used in the development and design of these structures must be based off of relevant and current rainfall data. Our findings of this work have proven that as the population of an area increases, and we build upon natural resources, the use and necessity of water resource management tools (including the development of storm water infrastructures) increases with the urbanization of that area. This information illustrates a key challenge for any community that wants to grow and develop while maintaining or improving their resilience to extreme rain events.

Acknowledgment: This work was supported by the National Science Foundation (NSF) and the Spelman College Summer Undergraduate Research (SURP) Program.

References

[1] Dhakal N., Jain S., Gray A., Dandy M., Stancioff E., Nonstationarity in seasonality of extreme precipitation: A nonparametric circular statistical approach and its application, *Water Resources Research*, 2015, 51(6), 4499-4515.
[2] Vogel R. M., Yaindl C., Walter M., Nonstationarity: Flood Magnification and Recurrence Reduction Factors in the United States, *Journal of the American Water Resources Association*, 2011, *47*(3), 464-474.
[3] Easterling D. R., Peterson T. C., Karl T. R., On the development and use of homogenized climate datasets, *Journal of Climate*, 1996, 9(6), 1429-1434.
[4] Jenkinson A. F., The frequency distribution of the annual maximum (or minimum) of meteorological elements, *Q. J. R. Meteorol. Soc.*, 1955, 81, 158-171.
[5] Coles S., *An introduction to statistical modeling of extreme values*, Springer, 2001.
[6] DeGaetano A. T., Time-dependent changes in extreme-precipitation return-period amounts in the continental United States, *Journal of Applied Meteorology and Climatology*, 2009, 48(10), 2086-2099.
[7] Mailhot A., Duchesne S., Design criteria of urban drainage infrastructures under climate change, *Journal of Water Resources Planning and Management*, 2009, 136(2), 201-208.
[8] Rust H. W., Kallache M., Schellnhuber H. J., Kropp J., *Confidence intervals for flood return level estimates using a bootstrap approach*, Springer, 2011, 61-81.

Danni Chen, Yuhuan Jia, Xiurui Guo*, Shuiyuan Cheng

Valuation of Health Benefit from the Air Quality Improvement in Beijing-Tianjin-Hebei and Surrounding Areas in 2020

Abstract: The Chinese government has defined the ambitious target of improving air quality during the 13rd FYP (Five-Year Plan). However, current knowledge of potential health benefits of air quality improvement remains insufficient. The paper aimed to evaluate the health impacts caused by the planning of air quality improvement in Beijing-Tianjin-Hebei (BTH) and surrounding areas by 2020, using the Poisson regression model. The results show that, the improvement in $PM_{2.5}$ levels in 2020 will result in substantial health benefits, for instance, 18,587 avoided premature deaths. The total economic value due to the decreased $PM_{2.5}$ concentration is 54.65 billion CNY, which is about 0.69% of GDP in 2013. And the health benefit is most massive in Beijing. The major health benefit is from the reduction of premature deaths and chronic bronchitis, which contribute to 88.63% of the total benefit. The results suggest that reduction of $PM_{2.5}$ concentration during the 13rd FYP in BTH and surrounding areas would benefit the public health substantially.

Keywords: air pollution control, health benefit, $PM_{2.5}$, BTH and surrounding areas.

1 Introduction

With the rapid economic growth in the past 30 years, China is suffering from severe air pollution, which caused an adverse impact on the health of residents, especially in mega-cities. It is proved that $PM_{2.5}$ is the representative air pollutant. A number of epidemiological studies have revealed that exposure to ambient $PM_{2.5}$ is associated with adverse health outcomes, especially mortality [1]. $PM_{2.5}$-related premature deaths throughout the world have increased from 800,000 in 2004 to 3.2 million in 2010, with Asia accounting for 72% in 2010 [2].

The Beijing-Tianjin-Hebei (BTH) region, one of the most developed and densely populated areas in China, is experiencing serious air pollution and health damage [3].

***Corresponding author: Xiurui Guo,** Key Laboratory of Beijing on Regional Air Pollution Control / College of Environmental & Energy Engineering, Beijing University of Technology, Beijing, 100124, China, E-mail: guoxiurui@bjut.edu.cn
Danni Chen, Yuhuan Jia, Shuiyuan Cheng, Key Laboratory of Beijing on Regional Air Pollution Control / College of Environmental & Energy Engineering, Beijing University of Technology, Beijing, 100124, China

Due to a large number of populations, even a slight decrease in $PM_{2.5}$ level in the future may lead to a substantial health benefit.

In order to address the severe particulate pollution, an action plan of Air Pollution Prevention and Control (APPC) was issued in 2013 [4], in which the target of specific $PM_{2.5}$ concentration was required to be achieved by 2017, such as the concentration in BTH region decreased by 25% compared to 2012. After that, MEP (Ministry of Environmental Protection of China) formulated the Implementation details of APPC in BTH and surrounding areas [5]. Considering that regional atmospheric pollution become increasingly serious, it would be more effective to take BTH and surrounding areas as a whole region to improve air quality. Related research [6] has shown that the transport among cross-city clusters outside BTH contributed 20%-35% of $PM_{2.5}$ emission in BTH. In fact, the deep mid- and long-term planning of air pollution control in BTH and surrounding areas is being drafted, in which the specific target of air quality improvement will be defined.

As we know, a critical and necessary component in the design and implementation of the air quality management policy is characterizing the impact on human health and the monetized benefits of air quality improvements [7]. The valuation of estimated health benefits would provide important information for the policymakers. Health impacts and benefits associated with air quality improvements have been studied by many researchers previously. Elena Boldo et al. estimated the benefits of implementing $PM_{2.5}$ control measures in Spain [8]. The results show that 1720 all-cause deaths in the over-30 age group and 1450 all-cause deaths in the 25–74 age group could be prevented with an average annual reduction of 0.7 $\mu g/m^3$ in $PM_{2.5}$ levels. Richard et al. found that reducing the $PM_{2.5}$ exposure in Sydney in 2007 by 10% (0.6 $\mu g/m^3$) would result in about 650 fewer premature deaths, a gain of 3500 life-years and about 700 fewer respiratory and cardiovascular hospital visits over 10 years [9]. An increasing number of Chinese researchers began to focus on this subject in recent years. Zou et al. [10] and Cheng et al. [11] analyzed the variation of PM_{10} concentration over time and related health impacts in Beijing and China, respectively. Lina et al. projected that the annual mean $PM_{2.5}$ concentration decreasing by 20.41 $\mu g/m^3$ in East China would generate at least 230 000 avoided deaths [12]. Ding et al. evaluated health benefit of emission control measures in Shanghai and found that the overall cost benefit of the improved air quality reached up to be 165 million CNY, with the avoided premature death contributing 90% of this figure [13].

This study aimed to obtain a better assessment regarding the health effects and the potential benefits of meeting the air quality targets in BTH and surrounding areas. The rest of this paper was organized as follows. In section 2, we introduced the methodology of our valuation and presented the data and variables. The results and discussions were presented in Section 3. We discussed the related uncertainties and limitations of this study in section 4. Finally, we drew our conclusions. The results would hopefully provide the impetus and scientific basis for relevant policy-making, and also enhance the public awareness of the health risk.

2 Methodology and Data

2.1 Study Area

The study area in this paper was defined as the BTH and surrounding areas, including Beijing, Tianjin and Hebei, Shanxi, Shandong, Henan and part of Inner-Mongolia. The BTH region is one of the most developed and densely populated areas in China. The Implementation details of APPC in 2013 defined BTH region and surrounding area as the focus of air pollution control in Northern China. The deep mid- and long-term planning of air pollution control in BTH and surrounding areas also defined this area as the foremost area to improve the air quality.

2.2 Method of Valuing the Health Benefit

1. *Assessment of health effects:* Based on the relative risk (RR) model in the form of a Poisson regression from the epidemiological studies, the log-linear exposure-response function below has been widely used to estimate the health impacts of air pollution [14].

$$I = I_0 \cdot exp[\beta \cdot (C - C_0)] \tag{1}$$

$$\Delta I = I - I_0 = I \cdot \{1 - 1/exp[\beta \cdot (C - C_0)]\} \tag{2}$$

where I is the mortality (morbidity) rate of the population exposed in the control year situation with concentration C and I_0 is the mortality (morbidity) rate of the population exposed in the baseline year situation with concentration C_0. β is the coefficient of association between air pollution concentration and mortality (morbidity). By multiplying the lessened mortality (morbidity) rate (ΔI) with the population, the number of avoidable deaths (diseases) associated with improved air quality can be calculated.

2. *Monetization of health effects:* Monetized health effects provide quantified economic benefits to policy-makers for evaluating air pollution control strategies. Using the unit value for each health endpoint, the reduced health impact can be monetized to the health benefit as follows:

$$TV = \sum V_i \cdot E_i \tag{3}$$

where E_i is the impact for health endpoint i, V_i is the unit economic value of health endpoint i, and TV is the sum of economic change of health endpoints.

Three monetization methods including willingness to pay (WTP), cost of illness (COI), and human capital (HC) approach are commonly used in valuating environmental health. In general, WTP is the most widely preferred used method because it takes intangible losses into account, such as pain, suffering and other adverse effects due to illness [15].

In this case, we calculated benefits of avoidable premature death using the VSL (value of statistical life) method, which represents the sum of what individuals would pay for reductions in their risk of dying that sum to saving one statistical life [16]. The unit value is generated from the original value for studies in reported city (Beijing) multiplied by the annual disposable income ratio per capita between Beijing and the source city [17]. Then the unit value for various currency years is adjusted to the year 2013 by multiplying by the annual consumer price index (CPI) in China.

For other health endpoints, we measure the benefits using the COI approach, formula is as follows:

$$C_i = C_{pi} + GDP_p \times T_{Li} \tag{4}$$

where C_i is the unit of health benefits of health endpoint i, C_{pi} is medical resources caused by health endpoints i, GDP_p is the daily mean value of per capita GDP, T_{Li} is the loss of working time.

2.3 Data

1. *Improved $PM_{2.5}$ level in 2020:* We collected the mean yearly $PM_{2.5}$ concentration in the base year of 2013 from Environment Quality Bulletin of each city. According to Ecological Environmental Protection Planning for Coordinated Development of BTH [18] and the 13rd FYP of each city or province, the mean $PM_{2.5}$ target concentration of each city in 2020 was designed. We assumed that the mean $PM_{2.5}$ concentration will be controlled within 55 μ/m³ in Beijing and Tianjin, about 60 μg/m³ in Shijiazhuang, Tangshan, Langfang, Jinan and Zhengzhou, about 70 μ/m³ in Baoding, Hengshui and Xingtai. And we also assumed that $PM_{2.5}$ concentration in Zhangjiakou, Chengde and Hohhot be controlled within the present GradeIIannual $PM_{2.5}$ air quality standard (GB 3095-2012).
2. *Population projection:* There are many factors influencing the urban population scale, and some of the factors are not completely sure. Grey models prediction method can forecast the system containing known information and uncertain factors. By filtering random quantity, the Grey model turns the raw data into orderly sequence and analyzes the regularity of changes through time. Because of its less information requirement and higher precision, the Grey model is suitable for complex and difficult prediction. In this case, we use GM (1, 1) to project the population in 2020.
3. *Selection of ERCs and mortality/morbidity rates:* The health endpoints were selected based on the literature [19]: (1) those registered in Chinese cities and classified by ICD-10 (International Classification of Diseases) code; (2) those published in exposure-response studies; and (3) statistical data such as mortality/

morbidity incidence rates. Accordingly, we selected the health endpoints of mortality for respiratory and cardiovascular disease, hospital admissions for respiratory and cardiovascular disease, pediatric and internal outpatient visits, chronic nd acute bronchitis, and asthma attack. As exposure-reaction coefficients (ERCs) establish the quantified relationship between the air pollution and health effect, selecting reasonable exposure-reaction coefficients is a key link in health effect assessment. In this case, those ERCs (Table 1) were selected through a comprehensive literature search to assess health impact. The annual incidence rates for each chosen health endpoint (in 2013) were obtained from the China health and family planning yearbook 2014 [20].

Table 1: ERCs and Annual Baseline Incidence Rates for Each of Health Endpoints

Health endpoints	β (95% CI)	Incidence rate in 2013
Mortality, cardiovascular	0.00053[21] (0.00015,0.0009)	0.0028[20]
Mortality, respiratory	0.00143[21] (0.00085,0.00201)	0.00076[20]
Hospital admission, cardiovascular	0.00068[21] (0.00043, 0.00093)	0.02[20]
Hospital admission, respiratory	0.00109[22] (0, 0.00221)	0.0099[20]
Outpatient visits, pediatric	0.00056[22] (0.0002, 0.0009)	0.12[20]
Outpatient visits, internal	0.00049[22] (0.00027, 0.0007)	0.28[20]
Chronic bronchitis	0.0075[22] (0.00025, 0.0135)	0.00148[19]
Acute bronchitis	0.0079[22] (0.0027, 0.013)	0.038[22]
Asthma attack	0.0021[22] (0.00145, 0.00274)	0.0094[22]

3 Results and Discussions

3.1 Physical Health Impacts

Table 2 listed the results of avoided health impacts from the strict $PM_{2.5}$ quality targets in BTH and surrounding areas in 2020. We conducted a statistical uncertainty analysis, but it is not comprehensive because we only calculated the 95% confidence interval for each health endpoint.

Table 2: The Health Impact from the Decrease of $PM_{2.5}$ Concentration in BTH and Surrounding Areas in 2020

Health endpoints	Health impact (thousand case)	95% CI
Mortality, cardiovascular	10907	3122-18317
Mortality, respiratory	7680	4645-10611
Hospital admission, cardiovascular	96678	61588-131252
Hospital admission, respiratory	78046	0-153108
Outpatient visits, pediatric	550739	198797-876312
Outpatient visits, internal	1132726	628135-1608423
Chronic bronchitis	67799	2778-104568
Acute bronchitis	1814050	716618-2617559
Asthma attack	140305	98746-179687

Improvement in $PM_{2.5}$ levels in 2020 was estimated to result in 18,587 avoided premature deaths, 1.75×10^5 avoided hospital admissions for cardiovascular and respiratory disease, 1.68×10^6 avoided outpatient visits, 67,799 avoided chronic bronchitis, 1.81×10^6 avoided acute bronchitis, and 1.40×10^5 fewer cases for asthma attack, across the BTH and surrounding areas. One can see from the table that avoided deaths are much fewer than hospital admission cases and outpatient visits. Given the same population in 2020 we projected previously, the main reason for this difference lies in the diverse ERCs and the incidence rate in calculation.

Health impacts of each city are summarized in Fig. 1. Obviously, health effects due to improvements in $PM_{2.5}$ levels in Shijiazhuang, Beijing and Tianjin are larger than that in other cities. One reason is the rapid urbanization that these cities have experienced in recent decades, which led to a constantly growing population exposed to and affected by $PM_{2.5}$ pollution. Another reason is the extent of $PM_{2.5}$ concentration reduction. Compared to 2013, the $PM_{2.5}$ concentration in Shijiazhuang, Beijing and Tianjin have decreased by 59.6%, 38.5% and 44.8%, respectively, relatively higher than other cities.

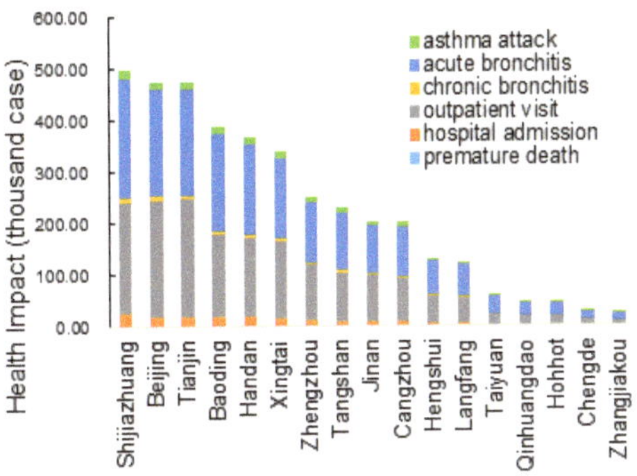

Figure 1: Health impact from air quality improvement in BTH and surrounding areas

3.2 Economic Valuation of Physical Health Impacts

Combining the health impacts we calculated above and the unit economic value of each health endpoint, we estimated the health benefits of improved $PM_{2.5}$ levels (Table 3). Based on this strategy, the total economic value due to the decreased $PM_{2.5}$ concentration was estimated 54.65 billion CNY, about 0.69% of GDP in 2013. The health benefits were mainly from the reduction in premature deaths and chronic bronchitis, which contributed to 88.63% of the total benefit. Although the avoided cases of premature deaths and chronic bronchitis are much fewer than other health endpoints, their much higher unit values result in the larger benefits.

Table 3: The Economic Benefit from the Decrease of $PM_{2.5}$ Concentration in BTH and Surrounding Areas in 2020

Health endpoints	Benefit (million CNY)	95% CI
Mortality, cardiovascular	13066.33	3737.13-21963.24
Mortality, respiratory	9020.62	5449.29-12477.74
Hospital admission, cardiovascular	1179.16	750.73-1601.78
Hospital admission, respiratory	928.68	0.00-1826.36
Outpatient visits, pediatric	208.09	75.05-331.37
Outpatient visits, internal	439.01	243.31-623.71
Chronic bronchitis	26348.57	1063.00-41083.43
Acute bronchitis	3277.20	1280.73-4772.87
Asthma attack	184.33	129.54-236.40

Health benefits of each city are summarized in Fig. 2. Obviously, Beijing gained the largest economic benefits, which reached up to 10.53 billion CNY. The health benefit of Beijing is followed by that of Tianjin and Shijiazhuang, which account for 8.17 billion and 6.50 billion, respectively. This reflects that we can gain enormous health benefits if we implement strict air pollution control measures in the BTH region.

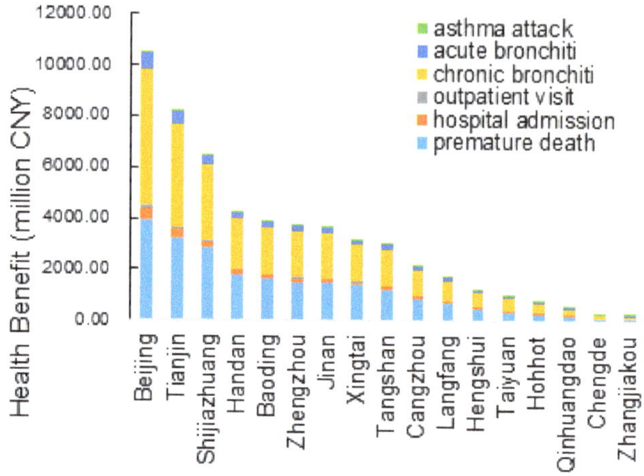

Figure 2: Health benefit from air quality improvement of each city in BTH and surrounding areas

We also calculated the GDP proportion and health benefits per capita of each city (Fig. 3). The highest estimated economic gain of health impacts is expected to be 618.67 CNY per capita in Shijiazhuang, which is about 1.34% of GDP in 2013. The second is Jinan, where the estimated economic gain is expected to be 520.95 CNY per capita and about 0.7% of GDP in 2013.

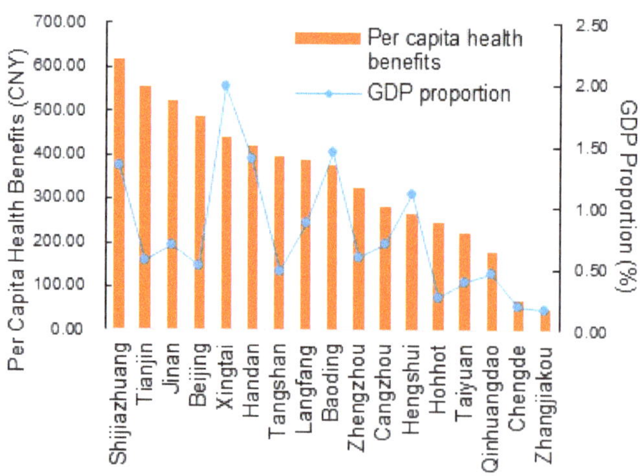

Figure 3. Per capita health benefits and GDP proportion of each city in BTH and surrounding areas

In summary, these numerical estimations indicated that improving air quality will have a positive impact on residents' health and the development of social economy. Our results are consistent with other studies in current literature. A Guangzhou study [13] found that by decreasing $PM_{2.5}$ concentration by 3.5 μg/m^3, health benefit was estimated to be 165 million CNY, 0.02% of GDP in Guangzhou in 2010. Our result seemed a little bit higher than it, since the improvement of $PM_{2.5}$ level was more significant in our study. Lina et al. [12] conducted a study in East China found that at least 230,000 deaths could be avoided by a large reduction in $PM_{2.5}$ concentration (20.41 μg/m^3) by 2030. Another study [23] conducted in China found that a control investment of 101.8 billion USD (0.79% of GDP) and a gain of 1.17% of China's GDP from improving $PM_{2.5}$ pollution are projected in 2030. Though these studies had different reduction of $PM_{2.5}$ concentration, population sizes, and characteristics, they all suggested that significant health benefits could be gained by effective air pollution control measures.

4 Uncertainties and Limitations

The data of this case study included air quality data, exposed population, ERCs, mortality/morbidity incidence rates, and monetary valuation functions. Each of these data can affect the final results to different extents. The accuracy of results is interpreted cautiously in the following discussion.

Air quality data and mortality/morbidity incidence rates were obtained from reliable data sources-the national/local statistics institute. In the process of population projection, we use the Grey models, which can accurately predict the population over the next five years, relatively. The validity of health impact results depends basically on the selection of ERCs. Compared with the foreign literature, there are few studies on the relationship between $PM_{2.5}$ and health impact for Chinese cities. This study selected domestic ERCs and considered regional differences as much as possible. Meanwhile, in the evaluation process, the 95% CI of health outcomes were used to reflect the error range.

Furthermore, uncertainty exists when using different unit values for monetization. The domestic studies and statistical data on the unit economic value for each health endpoint are relatively insufficient. Considering unit value varies greatly in the different regions and is generated by different monetization methods, we try our best to achieve accurate measurements. In this case, we select VSL to measure health benefits of avoided premature deaths, which can reflect people's willingness. And we obtain the unit value for each city in BTH and surrounding areas by adjusting result of Beijing with the annual capita disposable income

5 Conclusions

In summary, we can conclude that a reduction of $PM_{2.5}$ concentration in the BTH and surrounding areas could cause significant health benefit. Improvements in $PM_{2.5}$ levels due to the planning were estimated to result in 18,587 avoided premature deaths, 1.75×10^5 avoided hospital admissions for cardiovascular and respiratory disease, 1.68×10^6 avoided outpatient visits, 67,799 avoided chronic bronchitis, 1.81×10^6 avoided acute bronchitis, and 1.40×10^5 fewer cases for asthma attack, across the BTH and surrounding areas. Based on this estimation strategy, the total economic value due to decreased $PM_{2.5}$ concentration is 54.65 billion CNY, which is about 0.69% of GDP in 2013. The major health benefit is from the reduction in premature deaths and chronic bronchitis, which contributed to 88.63% of the total benefit.

Acknowledgment: This study was supported by the Natural Sciences Foundation of Beijing (Project No. 9152001) and the Ministry of Environmental Protection Special Funds for Scientific Research on Public Causes (No. 201409007). Many thanks for the anonymous reviewers for their valuable comments.

References

[1] Hoek G., Krishnan M.R., Beelen R., Peters A., Ostro B., Brunekreef B., et al. Long-term air pollution exposure and cardio-respiratory mortality: a review[J]. Environ. Health,2013, 12: 43.

[2] Apte J.S., Marshall J.D., Cohen A.J., Brauer M. Addressing global mortality from ambient $PM_{2.5}$[J]. Environ. Sci. Technol.,2015, 49: 8057–8066.

[3] Zhang L., Chen X., Xue X., Sun M., Han B., Li C., et al. Long-Term Exposure to High Particulate Matter Pollution and Cardiovascular Mortality: A 12-Year Cohort Study in Four Cities in Northern China[J]. Environ. Int.,2014, 62: 41–7.

[4] State Council, The notice on the action plan of the Air Pollution Prevention and Control from the State Council (in Chinese). http://www.gov.cn/zwgk/201309/12/con-tent_2486773.htm. Accessed 12 March 2016.

[5] Ministry of Environmental Protection of China, Implementation details of the action plan of Air Pollution Prevention and Control (APPC) in Beijing-Tianjin-Hebei and surrounding areas. Ministry of Environmental Protec¬tion of China (in Chinese). http://www.mep.gov.cn/gkml/hbb/bwj/201309/t20130918_260414.htm. Accessed 12 March 2016.

[6] Wang Z., Li J., Wang Z., Modeling study of regional severe hazes over Mid-Eastern China in January 2013 and its implications on pollution prevention and control[J]. Science China: Earth Sciences, 2014,57: 3–13, (in Chinese).

[7] National Academy of Sciences (NAS), Improving health in the United States: the role of health impact assessment[M]. Washington, DC: The National Academies Press, 2011.

[8] Boldo E., Linares C., Health impact assessment of a reduction in ambient $PM_{2.5}$ levels in Spain[J]. Environment International, 2011, 37: 342-348.

[9] Broome R.A., Fann N., The health benefits of reducing air pollution in Sydney, Australia [J]. Environmental Research, 2015, 143:19-25.

[10] Zou W., Zhang S., Economic Valuation of Health Impact of PM10 Pollution in Beijing from 2001 to 2006[J]. Chinese Journal of Population, Resources and Environment,2010, 2: 68-74.

[11] Cheng Z., Characteristics and health impacts of particulate matter pollution in China (2001-2011) [J]. Atmospheric Environment, 2013,65:186-194.
[12] Madaniyazi L., Projecting Fine Particu-late Matter-Related Mortality in East China[J]. Environ. Sci. Technol.2015, 49: 11141–11150.
[13] Ding D., Evaluation of health benefit using BenMAP-CE with an integrated scheme of model and monitor data during Guangzhou Asian Games[J]. Environ. Sci., http://dx.doi.org/10.1016/j.jes.2015.06.003
[14] Gao J., Yuan Z., Liu X., Xia X., Huang X., Dong Z. Improving air pollution control policy in China-a perspective based on cost-benefit analysis[J]. Sci. Total Environ.,2016,543: 307–314.
[15] Robinson L.A., (2011) Valuing the health impacts of air emissions. In: Nriagu, J.O. (Ed.), Encyclopedia of Environmental Health[M], Burlington: Elsevier, 2011: 619–627.
[16] Pope C.A. III, Cropper M., Coggins J.& Cohen A. Health benefits of air pollution abatement policy: Role of the shape of the concentration–response function[J]. Journal of the Air & Waste Management Association,2015, 65:5, 516-522.
[17] Huang D., Xu J., Zhang S., Valuing the health risks of particulate air pollution in the Pearl River Delta, China[J]. Environ. Sci. Pol.,2012, 15: 38–47.
[18] Ministry of Environmental Protection of China, Ecological Environmental Protection Planning for Coordinated Development of Beijing-Tianjin-Hebei region. Ministry of Environmental Protection of China (in Chinese). http://www.gov.cn/xinwen/201512/30/content_50 29650.htm. Accessed 20 May 2016.
[19] World Bank, China SEPA, Cost of pollution in China. Economic estimates of physical damages. http://siteresources.worldbank.org/IN TEAPREGTOPENVIRONMENT/Resources/China_Cost_of_Pollution.pdf. Accessed 31 March 2016.
[20] National Health and Family Planning Commission of China, China health and family planning yearbook 2014. The Minister of Health of China, Beijing. http://tongji.cnki.net.libproxy.bjut.edu.cn/kns55/Navi/Ye arBook.aspx?id=N2014120147&floor=1. Accessed 22 March 2016.
[21] Xie Y., Chen J., Li W., An Assessment of $PM_{2.5}$ Related Health Risks and Impaired Values of Beijing Residents in a Consecutive High-Level Exposure during Heavy Haze Days[J]. Environmental Science.,2014, 35(1): 1-8.
[22] Huang D., Zhang S.,Health benefit evaluation for $PM_{2.5}$ pollution control in Beijing-Tianjin-Hebei region of china[J]. China Environmental Science,2013, 33(1): 166-174.
[23] Xie Y., Dai H., Dong H., Economic Impacts from $PM_{2.5}$ Pollution-Related Health Effects in China: A Provincial-Level Analysis[J]. Environ. Sci. Technol., 2016, 50: 4836–4843.

Shimin Zhai, Xuesong Guo, Rongzhan Liu*, Benyi Xiao

Enhancement Cationic Dyes Adsorption of Sewage Sludge-based Bio-chars from Solutions by Thermal-alkaline Pretreatment

Abstract: In this study, sewage sludge based bio-char (SBC) and thermal-alkaline treated sludge based bio-char (TSBC) were used to adsorb cationic red X-GRL and cationic yellow X-6G. The effect of bio-char dosage, adsorption kinetics and adsorption isotherm experiment were also studied. The adsorbed amount of cationic dyes onto TSBC and SBC were 45-50 mg g^{-1} and less than 11 mg g^{-1} at room temperature, respectively. The results showed that pseudo-second order kinetics was the most suitable model for describing the adsorption of cationic dyes onto bio-char and equilibrium data were well fitted to the Freundlich isotherm model. The adsorption mechanism between cationic dye and bio-char involved electrostatic interaction, ion exchange, surface complexation, physical function and others. As sewage sludge-based bio-char can get more pore structures by thermal-alkaline pretreatment, TSBC has higher capabilities of removing cationic dyes than SBC.

Keywords: bio-char, cationic dye, adsorption, sewage sludge, enhancement.

1 Introduction

Dyes and their effluent have become one of the main sources of water pollution following the development of printing and dyeing industry [1]. Cationic dyes are mainly applied to acrylic fibers and their effluent is one of the most difficult wastewater to treat because of their complex composition, high salinity, low pH value and bad biodegradation. At present, the main treatments of cationic dye wastewater are chemical coagulation, adsorption, oxidation method and membrane separation method [2]. Adsorption is a common treatment process to remove cationic dye from wastewater, which use the porous surface reactive solid contact with the wastewater and concentrates certain dyes on the solid surface from liquid.

Adsorption method was considered as the better treatment of cationic dye wastewater based on its higher efficiency, simple operation and insensitivity to toxic substances, etc. The traditional absorbents include commercial activated carbon,

***Corresponding author: Rongzhan Liu,** College of Textiles and Clothing, Qingdao University, Qingdao, China, E-mail: lrz760504@126.com
Shimin Zhai, College of Textiles and Clothing, Qingdao University, Qingdao, China
Xuesong Guo, Benyi Xiao, Research Center for Eco-Environmental Science, Chinese Academy of Sciences, Beijing, China

resins, natural mineral materials, and so on [3]. However, their high costs inhibit their extensive application. Therefore, the study of new and cheap materials is viewed as more and more important [4]. The biomass materials, which are renewable, cost-effective and obtained easily from the environment, are recently gaining more and more attention.

Meanwhile, sewage sludge has become a stringent problem in wastewater treatment plants. As sewage sludge contains lots of heavy metals, bacteria, fungi and organic materials, it will create secondary pollution if its treatment and disposal is not suitable [5]. Incineration, composting, landfill, anaerobic digestion and soil utilization were commonly used treatment methods. The fundamental outlet to sludge disposal is resource utilization, which leads to many researches currently, such as, bio-char, hydrogen production and compost etc. [6]. Bio-char preparation from sewage sludge is one of the resource utilization methods. Sewage sludge can be prepared by adsorption materials as lots of organisms are contained by it. Previous research showed that sludge based bio-char has good capability of adsorption because of its large specific surface and high porosity [7]. But the adsorption capability of sludge based bio-char is usually less, and the micropore size distribution is generally narrower than that of commercial activated carbon [8]. How to enhance the adsorption capability of sludge based bio-char is a problem that urgently needs to be solved [9].

In this research, sludge bio-char (SBC) and thermal-alkaline treated sludge bio-char (TSBC) were prepared from municipal sewage sludge and thermal-alkaline treated sludge. The SBC and TSBC were applied to the adsorption of cationic red X-GRL and cationic yellow X-6G from solutions. The purpose of this research was to enhance the adsorption effect of cationic dyes onto sludge based bio-char by thermal-alkaline pretreatment, study the effect of bio-char dosage and adsorption time on cationic dye adsorption, research the kinetics of cationic dye adsorption on two kinds of bio-char and explore the mechanism of cationic dye adsorption on bio-char.

2 Materials and Methods

2.1 Sewage Sludge and Its Thermal-alkaline Pretreatment

Sewage sludge (SS) was collected from a wastewater treatment plant in Beijing (China), which handled 60,000t wastewaters daily by anaerobic-anoxic-oxic process. The pH value of the sludge samples was about 6.9±0.1. The total chemical oxygen demand (TCOD) of the samples was 11,500 mg/L. The soluble chemical oxygen demand (SCOD) was 177.3 mg/L. The collected sludge samples were gravitationally settled for about 1 h and the sediments were stored at 4°C before being used. The pH of settled sludge was adjusted to 13.0 by adding 6 M NaOH, and then the sludge was alkaline pretreated under 175°C for 1 h. The pretreated sludge was the sediments of thermal-alkaline pretreated sludge (TAS).

2.2 Preparation of TSBC and SBC

The sewage sludge and thermal-alkaline pretreated sludge were dried for 24 h at 105°C, and then crushed into 0.15 mm particle size. SBC and TSBC were prepared from SS and TAS at 600°C for 1 h under nitrogen environment. The pyrolysis of the dried SS and TAS were performed in a horizontal quartz reactor placed in a laboratory electrical furnace. The bio-char was crushed into 0.15 mm particle size.

2.3 Cationic Dyes

Cationic dyes are often divided into two categories: isolated-type cationic dyes and conjugated-type cationic dyes [10]. The positive charge of isolated-type cationic dyes is concentrated. But the positive charge of conjugated-type cationic dyes is distributed on the conjugated systems [11]. To study the effect of positive charge distribution on cationic dye adsorption, cationic red X-GRL (isolated-type) and cationic yellow X-6G (conjugated-type) were tested in this study. Their properties are shown in Table 1.

Table 1: Properties of Selected Cationic Dyes

Name	Category	Molecular structure	Molecul-ar formula	Molecular weight
Cationic red X-GRL	Isolated-type	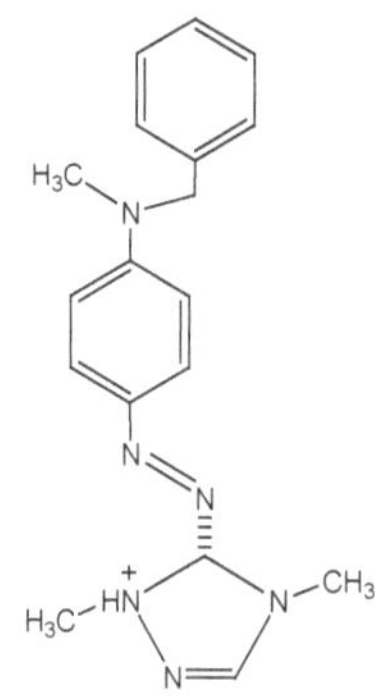	$C_{18}H_{23}N_6$	323
Cationic yellow X-6G	Conjugated-type	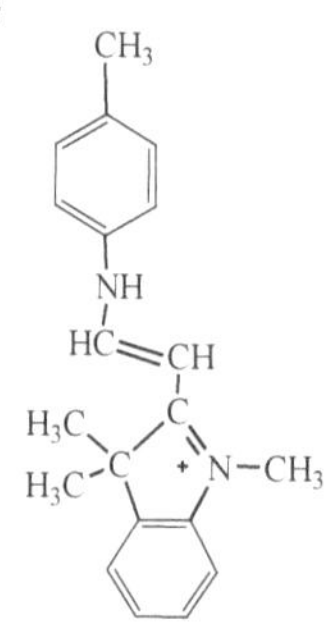	$C_{20}H_{23}N_2$	291

2.4 Cationic Dyes Adsorption

The effect of bio-char dosage was performed in 20 mL of cationic red X-GRL and cationic yellow X-6G solution of 40 mg/L respectively. Different weights of SBC (0.02 g - 0.2 g) and TSBC (0.01 g - 0.1 g) were added to dye solution and stirred at 25°C for 24 h. The concentrations of dyes left in solution were measured respectively and the removal efficiency and amount adsorbed of two kinds of dye were calculated. The experiments were duplicated and only the average values were reported. The removal efficiency and adsorption amount were calculated.

Based on the above experimental parameters, a fixed amount of SBC and TSBC were added to 20 mL of cationic red X-GRL and cationic yellow X-6G solution of 40 mg/L respectively. The bio-char and dye solutions were stirred at 25°C for different time. Then, the concentrations of dyes left in solution were measured and the amount of two kinds of dye adsorbed, was calculated.

The adsorption data were fitted by the pseudo-first-order and pseudo-second-order kinetics equation [12]. The pseudo-first-order kinetic equation was shown as follows:

$$\ln(q_e - q_t) = \ln q_e - K_1 t \tag{1}$$

In (1), qe is the adsorption capability at equilibrium time (mg g^{-1}); qt is the adsorption capability at time t (mg g^{-1}); K1 is the pseudo-first-order rate constant (min^{-1}). The pseudo-second-order kinetic equation was:

$$\frac{t}{q_t} = \frac{1}{K_2 q_e^2} + \frac{t}{q_e} \tag{2}$$

In (2), qe is the adsorption capability at equilibrium time (mg g^{-1}); qt is the adsorption capability at time t (mg g^{-1}); K1 is the pseudo-second-order rate constant (min^{-1}).

Based on the above experimental parameters, a fixed amount of SBC and TSBC were added to 20 mL of cationic red X-GRL and cationic yellow X-6G solution of different concentrations respectively. Freundlich and Langmuir isotherm models were used to describe the adsorption process. Freundlich equation was [13-14]:

$$Q_e = K_F C_e^{1/n} \tag{3}$$

Langmuir equation was:

$$Q_e = \frac{q_m K_1 C_e}{1 + K_L C_e} \tag{4}$$

In (3) and (4), Qe is the adsorption capacity at equilibrium time (mg g^{-1}); Ce is the equilibrium concentration of dye solution; KF is the Freundlich constant (L mg^{-1}); 1/n is the heterogeneity of the sorption sites and an indicator of isotherm nonlinearity; KL is the Langmuir constant (L mg^{-1}); qm is the ideal maximum monolayer adsorption capability (mg g^{-1}).

2.5 Analytical Methods

UV-spectrophotometer was used to establish a standard curve with absorbance and cationic dye concentration at maximum wavelength. The bio-char samples before and after cationic dye adsorption were analyzed using a SEM (S-4800, Hitachi, Japan). The Fourier Transform-infrared (FTIR) spectrum was measured on a Nicolette is50 FTIR spectrometer (Thermo Fourier, USA).

3 Results and Discussion

3.1 Effect of Bio-char Dosage

The effect of TSBC and SBC dosage on cationic dye adsorption was shown as Fig. 1. In Fig. 1, when the dosage of TSBC was more than 2 g/L, the removal rate of cationic yellow X-6G could reach above 98% while the least dosage of SBC was 6 g L^{-1}. When the dosage of TSBC addition reached 1.5 g L^{-1}, the removal rate of cationic red X-GRL was 99% or above and the same result needed the dosage of SBC addition to reach 5 g L^{-1}. It indicated that the dosage of bio-char addition was a significant factor influencing cationic dye adsorption. The TSBC and SBC were effective to decolorize cationic dyes and the TSBC have higher adsorption capacities.

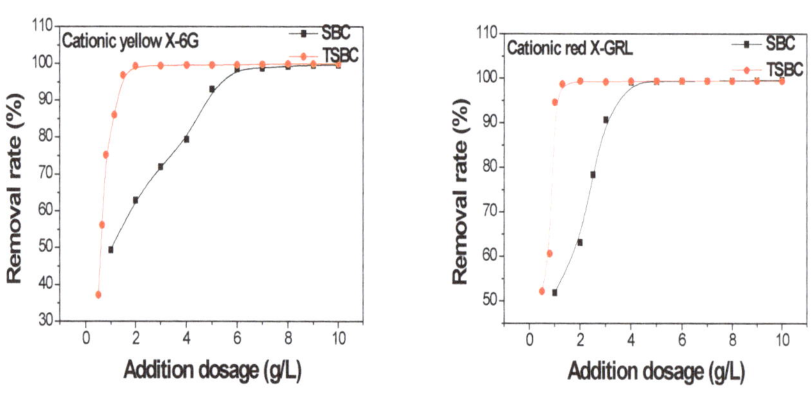

Figure 1: Effect of bio-char dosage on cationic dye adsorption

The reason could be that TSBC prepared from thermal-alkaline treated sludge is strongly alkaline and can link with cationic dye through ionic bonds more easily than SBC [16]. In addition, it is advantageous to the formation of bio-char's porous structures with NaOH addition [17]. The rich porous structure gave TSBC higher adsorption capacities. Compared with cationic yellow X-6G, The removal effect of cationic red X-GRL is better. Because cationic red X-GRL is isolated-type, the positive charge is concentrated and can link with the negative charge of bio-char easily. But cationic yellow X-6G is conjugated-type, the positive charge is distributed on the conjugated systems [18]. The ionic bond force between cationic yellow X-6G and bio-char is weaker than that of cationic red X-GRL. The analysis is consistent with the SEM photograph shown in Fig. 1.

3.2 Adsorption Kinetics

In order to investigate the adsorption kinetics and mechanism of sludge based bio-char for two cationic dyes, pseudo-first and pseudo-second order equations were used to fit the kinetics experimental data and the results are shown in Fig. 2. It can be seen that the adsorption equilibrium amounts of cationic red X-GRL and yellow X-6G on the same bio-char were identical. The adsorption equilibrium amounts of SBC and TSBC were about 4.9 mg g^{1} and 19.9 mg g^{1} respectively. All of the adsorption equilibriums were reached after about 90min and the adsorption equilibrium amount of TSBC for cationic dyes was higher than that of SBC. Additionally, both pseudo first and second model can fit the adsorption process well before 90min and the adsorption kinetic data were fitted better by the pseudo second-order kinetic model than pseudo first equation after 90 min.

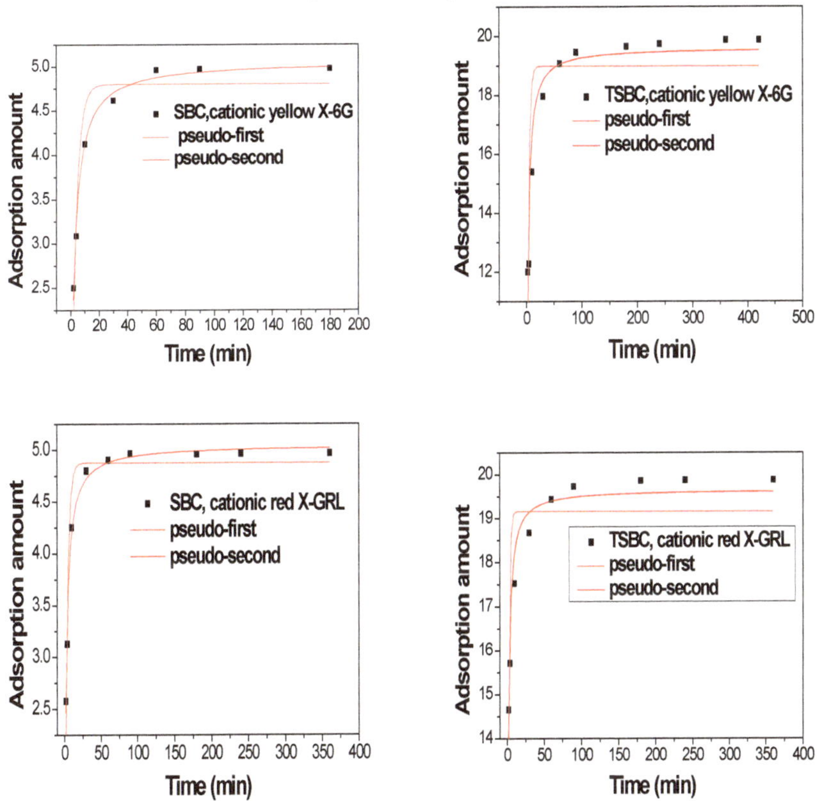

Figure 2: Kinetics fits for cationic dyes adsorption on bio-char

The fitting parameters of the pseudo-first-order and pseudo-second-order kinetics equations were shown in Table 4. It can be seen that the Qe values of pseudo-first-order kinetics equation were about 4.8 mg g^{1} (SBC) and 19 mg g^{-1} (TSBC). The Qe values of pseudo-second-order kinetics equation were about 5 mg g^{1} (SBC) and 19.6 mg g^{1} (TSBC), which were closer to the test results (about 4.9 mg g^{1} and 19.9 mg g^{1} respectively). In addition, the R^2 values of pseudo-second-order kinetics equation were above 0.9 and higher than that of pseudo-first-order kinetics equation. Above results suggest that pseudo-second-order model is more suitable to fit the dyes adsorption of two bio-chars.

Pseudo-first-order model is usually applied in physical adsorption process, while pseudo-second-order model is applied when chemical adsorption exists in the process [19]. Through the experiments made on the adsorption of TSBC and SBC on two types of cationic dyes, it is found that not only physical adsorption but also chemical adsorption were included in the process. The transfer and exchange of electrons were included in the adsorption process [20].

To study the adsorption isotherm process of cationic dye on bio-char, Langmuir and Freundlich isotherm models were used to describe the adsorption process. The fitting curves and fitting parameters are summarized in Fig. 3 and Table 3.

Table 2: Fitting Parameters of Kinectic Equations

Dyes	Bio-char	Pseudo-first-order kinetic equation			Pseudo-second-order kinetic equation		
		K_1 (h^{-1})	Qe (mg g^{-1})	R^2	K_2 (mg g^{-1}/h)	Qe (mg g^{-1})	R^2
Cationic yellow X-6G	SBC	0.285	4.804	0.904	0.086	5.062	0.989
	TSBC	0.332	18.992	0.713	0.027	19.623	0.917
Cationic red X-GRL	SBC	0.290	4.877	0.931	0.094	5.049	0.988
	TSBC	0.613	19.170	0.666	0.062	19.683	0.934

Table 3: Fitting Parameters of Isotherm Models

Dye	Bio-char	Langmuir models			Freundlich models		
		K_L(L/mg)	qm(mg/g)	R^2	KF($mg^{1/n}L^{1/n}$/g)	1/n	R^2
Cationic yellow X-6G	SB	15.578	10.222	0.95	7.743	0.06	0.80
	TSB	6.129	39.295	0.58	25.948	0.10	0.95
Cationic red X-GRL	SB	4.926	10.643	0.98	7.713	0.07	0.67
	TSB	4.472	41.735	0.76	27.532	0.09	0.92

3.3 Adsorption Isotherm

In Fig. 3, It can be seen that there are more points on Langmuir curve for the cationic dyes adsorption on SBC and more points were on Freundlich curve for the adsorption on TSBC. In other words, Freundlich models can describe the adsorption process of cationic dyes on TSBC better than Langmuir models and the adsorption process of cationic dyes on SBC can be described better by Langmuir models. In Table 4, the qm values of cationic dyes adsorption on SBC were 10.2 mg g^{1} and 10.6 mg g^{-1}, which were closer to the test results (10.6 mg g^{1} and 10.7 mg g^{1}). The R2 values of Langmuir models for adsorption on SBC were 0.95 and 0.98, which were higher than the values of TSBC (0.58 and 0.76). Additionally, it is obvious that the R2 values of Freundlich models for adsorption on TSBC were above 0.9. It shows that Freundlich models can fit the cationic dyes adsorption on TSBC well and Langmuir models can fit the cationic dyes adsorption on SBC well, which is consistent with the result in Fig. 3.

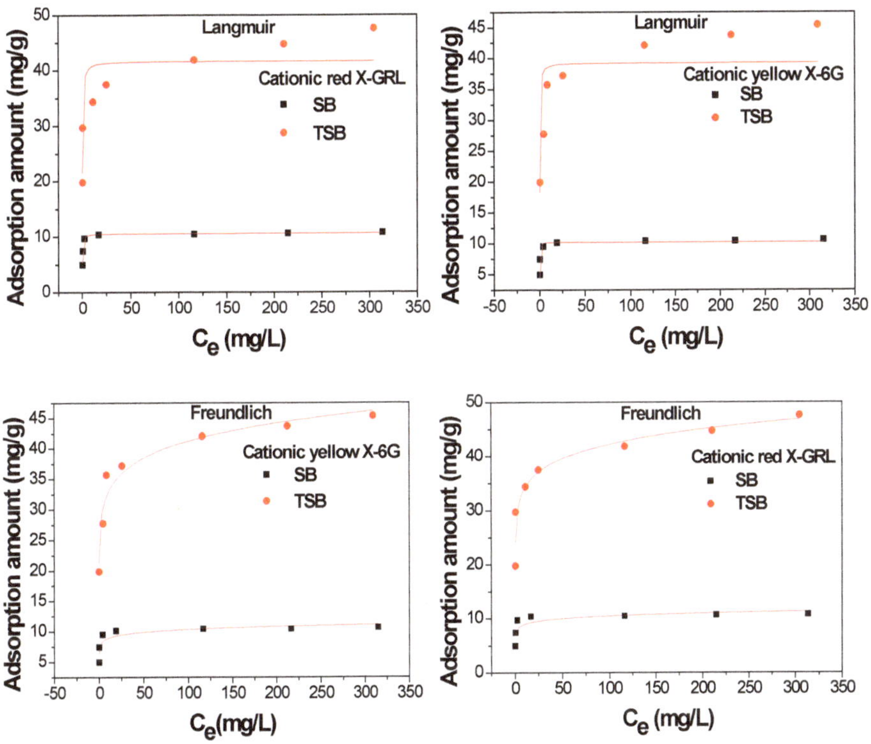

Figure 3: Curve fits of isotherm models for cationic dyes adsorption on bio-char

The analysis results mean that the adsorption on TSBC is not only single ply adsorption but also physical adsorption and chemical adsorption (include electrostatic attraction, complexation of functional groups and hydrogen bonding) [21-22]. But the Langmuir models can fit the adsorption process of cationic dye onto SBC well. The correlative data indicates that the adsorption data can be fitted well by the Langmuir models when the adsorbability is weak [23]. The weak adsorbability of SBC may be the reason why it is different from the fitting results of TSBC.

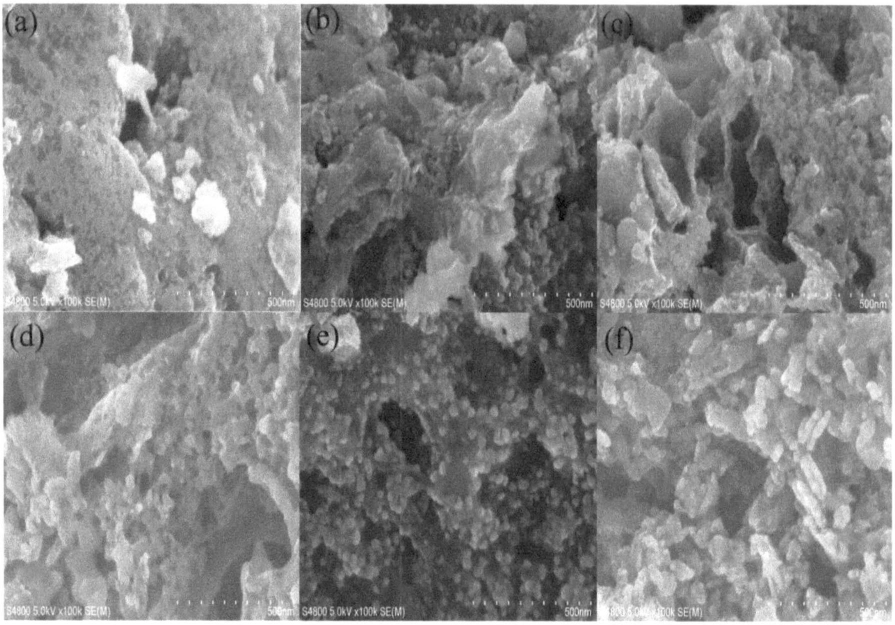

Figure 4: SEM photographs of bio-char: (a) SBC, (b) SBC with cationic red X-GRL, (c) SBC with cationic yellow X-6G, (d) TSBC, (e) TSBC with cationic red X-GRL adsorbed and (f) TSBC with cationic yellow X-6G adsorbed.

SEM photographs of bio-char and bio-char with cationic dye adsorbed are shown in Fig. 1. It can be seen that both TSBC (d) and SBC (a) have many pores, but the pore structure of TSBC (d) is more rich and significant, which suggested that thermal-alkaline treatment could enhance the formation of pores structure. Because the structure characteristic of bio-char affects its potential to an adsorbent [24], TSBC have higher adsorption capacities than SBC. In SEM photographs of bio-char after adsorption (b-f), it can be seen that some dye particles appeared on the surface or inside of bio-char. The phenomenon indicated that the cationic dyes were adsorbed by bio-char. In addition, the dye particles on TSBC surface were much more than those on the SBC surface obviously, which were consistent with the obtained results from Fig. 4.

3.4 FTIR Analysis

The FTIR analysis results of TSBC and SBC were shown as Fig. 5. It can be seen that the two kinds of bio-char have abundant functional groups. The functional groups of SBC and TSBC were mostly similar. In characteristic area, the infrared spectra include 3413 cm^{-1} peak (bonded -OH groups), 2927 cm^{-1} peak (aliphatic -C-H), 2858 cm^{-1} peak (aliphatic -C-H), 1668 cm^{-1} peak (C=O stretching), 1570 cm^{-1} peak (aromatic ring), 1379 cm^{-1} peak (C-H or $-CH_3$ bending), 1039 cm^{-1} peak (C=O groups) and 464 cm^{-1} peak (-N-H) [25].

The polar groups of bio-char (-OH, -COOH, $-NH_2$) can link with cationic dye through the strength of ionic bonds, Van der Waals force and hydrogen bonds easily [16]. These polar groups can play an important role in the adsorption or binding process. The mechanism between cationic dye and bio-char may involve electrostatic interaction, ion exchange, surface complexation, physical function and others [22].

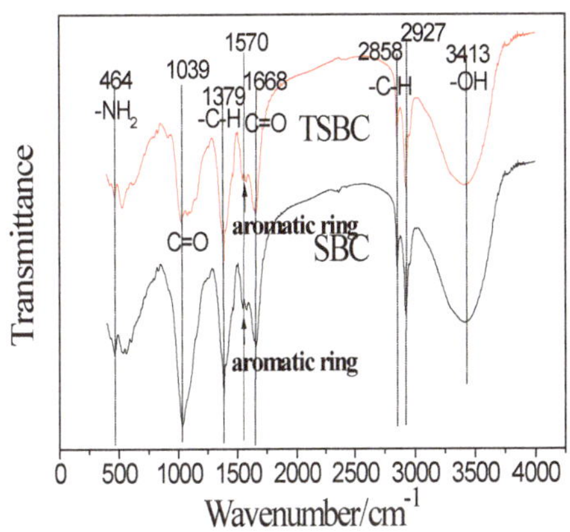

Figure 5: FTIR spectra of TSBC and SBC.

4 Conclusions

In this study, the adsorption process of cationic dye onto SBC and TSBC was explored. The result indicated that TSBC has a good decolorizing effect on cationic dye. The adsorption of cationic dyes onto bio-char conforms pseudo-second order kinetics model and equilibrium data were well fitted to the Freundlich isotherm model. The mechanism between cationic dye and bio-char involved electrostatic interaction, ion exchange, surface complexation, physical function and so on. Therefore, thermal-alkaline treatment could enhance the adsorbtion capability of sludge bio-char and thermal-alkaline treated sludge bio-char could be a kind of valuable absorbent to cationic dye in the wastewater.

References

[1] Fan S, Tang J, Wang Y, et al. Bio-char prepared from co-pyrolysis of municipal sewage sludge and tea waste for the adsorption of methylene blue from aqueous solutions: Kinetics, isotherm, thermodynamic and mechanism[J]. J. Mol. Liq., 2016, 220:432-441.

[2] Cheng G, Sun L, Jiao L, et al. Adsorption of methylene blue by residue bio-char from copyrolysis of dewatered sewage sludge and pine sawdust[J]. Desalin. Water Treat., 2013, 51(37-39):7081-7087.

[3] Himanshu Patel, R. T. Vashi. A comparison study of removal of methylene blue dye by adsorption on Neem leaf powder (NLP) and activated NLP[J]. J. Environ. Eng. Landsc., 2013, 21(21):36-41.

[4] Qiu M, Huang C. Removal of dyes from aqueous solution by activated carbon from sewage sludge of the municipal wastewater treatment plant[J]. Desalin. Water Treat., 2013, 53(13):1-8.

[5] Jusufi K, Halili J, Berisha A, et al. Removal of dyes from wastewater using plant-based biosorbent derivate from potato peels[C]. RAD, 2016.

[6] Yusoff I. Removal of acid yellow-17 dye from aqueous solution using eco-friendly bio-sorbent[J]. Desalin. Water Treat., 2013, 51(22):4530-4555.

[7] Li W H, Yue Q Y, Gao B Y, et al. Preparation and utilization of sludge-based activated carbon for the adsorption of dyes from aqueous solutions[J]. Chem. Eng. J., 2011, 171(1):320-327.

[8] Li W H, Yue Q Y, Gao B Y, et al. Preparation of sludge-based activated carbon made from paper mill sewage sludge by steam activation for dye wastewater treatment[J]. Desalination, 2011, 278(1–3):179-185.

[9] Martin M J, Artola A, Balaguer M D, et al. Activated carbons developed from surplus sewage sludge for the removal of dyes from dilute aqueous solutions[J]. Chem. Eng. J., 2003, 94(3):231-239.

[10] Sivarajasekar N, Baskar R. Adsorption of basic red 9 on activated waste Gossypium hirsutum, seeds: Process modeling, analysis and optimization using statistical design[J]. J. Ind. Eng. Chem., 2014, 20(5):2699-2709.

[11] Raj K R, Kardam A, Arora J K, et al. Adsorption behavior of dyes from aqueous solution using agricultural waste: modeling approach[J]. Clean Technol. Envir., 2013, 15(1):73-80.

[12] Păcurariu C, Paşka O, Ianoş R, et al. Effective removal of methylene blue from aqueous solution using a new magnetic iron oxide nanosorbent prepared by combustion synthesis[J]. Clean Technol. Envir., 2015, 18(3):705-715.

[13] Meng J, Feng X, Dai Z, et al. Adsorption characteristics of Cu(II) from aqueous solution onto bio-char derived from swine manure[J]. Envir. Sci. Pollut. R., 2014, 21(11):7035-46.

[14] Meng J, Wang L, Liu X, et al. Physicochemical properties of bio-char produced from aerobically composted swine manure and its potential use as an environmental amendment[J]. Bioresource Technol., 2013, 142(4):641-646.

[15] Dai Z, Wang Y, Muhammad N, et al. The Effects and Mechanisms of Soil Acidity Changes, following Incorporation of Bio-chars in Three Soils Differing in Initial pH. Soil Sci. Soc. Am. J., 2014, 78(5):1606-1614.

[16] Ma L, Rao X, Lu P, et al. Acid-tolerant plant species screened for rehabilitating acid mine drainage sites. J. Soil. Sediment., 2015, 15(5):1104-1112.

[17] Adeyemo A A, Adeoye I O, Bello O S. Adsorption of dyes using different types of clay: a review[J]. Appl. Water Sci.. 2015:1-26.

[18] Menkiti M C, Aneke M C, Ejikeme P M, et al. Adsorptive treatment of brewery effluent using activated Chrysophyllum albidium seed shell carbon[J]. Springerplus, 2014, 3(1):213-213.

[19] Akinpelu D A, Odewade J O, Aiyegoro O A, et al. Biocidal effects of stem bark extract of Chrysophyllum albidium, G. Don on vancomycin-resistant Staphylococcus aureus[J]. Bmc Complem. Altern. M., 2016, 16(1):1-9.
[20] Hu J, Yang X, Shao L, et al. Effect of Alkali Treatment on Heavy Metals Adsorption Capacity of Sewage Sludge[J]. J. Geosci. Envir. Prot., 2015, 03(2):33-39.
[21] Wang L, Butterly C R, Wang Y, et al. Effect of crop residue biochar on soil acidity amelioration in strongly acidic tea garden soils[J]. Soil Use. Manage., 2014, 30(1):119–128.
[22] Yuan J H, Xu R K. Effects of biochars generated from crop residues on chemical properties of acid soils from tropical and subtropical China[J]. Soil Res., 2012, 50(7):570-578.
[23] Zhao R, Coles N, Kong Z, et al. Effects of aged and fresh bio-chars on soil acidity under different incubation conditions[J]. Soil. Till. Res., 2015, 146(146):133-138.
[24] Zhang B, Xiong S, Bo X, et al. Mechanism of wet sewage sludge pyrolysis in a tubular furnace[J]. Int. J. Hydrogen Energ., 2011, 36(1):355-363.
[25] Liu L, Yan L, Liu Y, et al. Removal of Methylene Blue from Aqueous Solutions by Sewage Sludge Based Granular Activated Carbon: Adsorption Equilibrium, Kinetics, and Thermodynamics[J]. J. Chem. Eng. Data, 2013, 58(58):2248-2253.

Yawen Zeng*, Xiaoying Pu, Xiaomeng Yang, Jiazhen Yang, Juan Du, Tao Yang, Xia Li

Strategies of Functional Foods for Heart Disease Prevention in Human Beings

Abstract: Functional food is used for modifying the key risk factors for heart disease. To obtain a better understanding of functional food crop for heart disease prevention, we conducted a systematic review for risk factors and prevention strategies for heart disease using the PubMed and so on database for the period 2001-2016. Major heart disease risk factors of dietary except heredity in human are low dietary fiber, obesity and high elements intakes. The countermeasures of heart disease prevention in humans are the functional foods with brown rice, whole wheat, barley grain/grass powder and its product, oat and its product, vegetables, fruits, and mushrooms for dietary, which associated with food type of ancient human beings. These are the reasons that Western countries have more coronary heart disease than that of stroke and diabetes for Asian countries, based on the loss of K and Mg as well as dietary fiber of major food from whole wheat to wheat flour. We can suggest that chronic disease especially heart disease of human beings was associated the five evolutionary stages of the major dietary guidelines, which was as follows: I-fruits or vegetables, II-grass or Cyperaceou, III-cereals (rice, wheat, millet, beans, barley and corn), IV-polished rice or wheat flour, V-white rice or wheat flour + grass powder, i.e. the most healthy major dietary guidelines for modern humans.

Keywords: functional food, dietary, heart disease prevention, human being.

1 Introduction

Heart disease is the main cause of mortality world-wide, costing U.S $503 billion in 2010. Congenital heart disease (CHD) is a question for structure and function of heart at birth, however acquired heart diseases including the coronary (CHD), ischemic (IHD), hypertensive (HHD), rheumatic (RHD), pulmonary (PHD), hyperthyrod (HYHD),

***Corresponding author: Yawen Zeng,** Biotechnology and Genetic Resources Institute, Yunnan Academy of Agricultural Sciences / Agricultural Biotechnology Key Laboratory of Yunnan Province, Kunming 650205, Yunnan, China, E-mail: zengyw1967@126.com
Xiaoying Pu, Xiaomeng Yang, Juan Du, Tao Yang, Xia Li, Biotechnology and Genetic Resources Institute, Yunnan Academy of Agricultural Sciences / Agricultural Biotechnology Key Laboratory of Yunnan Province, Kunming 650205, Yunnan, China
Jiazhen Yang, Kuming Tiankang Science & Technology Limited Company,Kunming 650231, Yunnan, China

metabolic (MHD), infective etc. eight types heart disease. The depression is related with an increased risk of CHD and stroke [1]. More than 50% of global mortality rates of all heart diseases are preventable. Natural products play important roles against heart disease [2,3]. The diet and exercise can be very effective in reducing coronary risk, even outperforming drug therapy [4]. Liver X Receptor-a associated with CHD and neuro-degenerative diseases have been found to be regulated by the dietary components [5]. Children can develop habits which reduce their risk of heart disease, based on tobacco avoidance, healthy food choices, and regular physical activity [6]. The Mediterranean-style diet reduce greater cardiovascular diseases than low-fat diets, this diet including vegetables, fruit, whole grains, olive oil, and fish [7].

Functional foods are powerful tools for maintaining health and fighting against heart disease [8]. Their global market is about 73 billion euros with 8%–16% growth rate every year [9]. The main bioactive constituents for foods with heart disease prevention include angiotensin I, polyphenols (tea), flavonoids (buckwheat), flavanols (vegetables), cathecins (green tea), anthocyanins (blueberry), phenolic acids (fruits), tannins (plant), resveratrol, saponin (panax notoginseng), sterols (barley), vitamins, polysaccharides (mushrooms), fiber, as well as K, Ca, and P [10]. Functional foods were used as bone health, heart disease, and arthritis [11]. A diet with high long-chain n-3 fatty acids intake may prevent cardiovascular mortality associated with increased resting heart rate [12]. Modifiable risk factors for coronary artery disease (CAD) were associated with diet with low calorie intake [13]. This article was expounded by a wide range of functional foods for heart disease preventing in China and discuss reasons caused heart disease and strategies preventing.

2 Major Heart Disease Risk Factors for Dietarey

2.1 Heredity is a Major Factor that Causes Heart Disease

Genetic factors may play an important role for heart disease development in cases not related to environmental factors [14]. Inhibiting microRNA-34 and microRNA-25 benefit heart disease [15,16]. 16 genes encoding structural proteins and transcription factors, which are associated with congenital heart disease in humans [17]. The expression levels of TCF21 and miR-224 were disturbed in human atherosclerotic lesions, which revealed the complex heritable mechanisms of CHD risk [18]. The apoE ε4 allele is a risk factor for CHD in China which revealed the complex heritable mechanisms of CHD risk [19].

NKX2-5 gene is associated with a small number of ventricular septal defect, and rs2277923 SNP is associated with the risk of sporadic atrial septal defect Yunnan population in China [20]. Tumor necrosis factor alpha 308G>A associated with the increased risk of RHD [21]. Homocysteine is an important risk factor for lots of cardiovascular diseases, but its removal is very important in cardiac development

[22]. A comprehensive list of 75 single nucleotide variations have been collected, and predisposed individuals associated with cardiac arrhythmias [23]. Four loci for CAD located in or near TTC32-WDR35, GUCY1A3, C6orf10-BTNL2 and ATP2B1, which are associated with the susceptibility for CAD of Han population in China [24]. 12 SNPs are associated with a lipid trait, 5 SNPs are associated with blood pressure, meanwhile, four most important pathways are linked to lipid metabolism and inflammation and so on activities in the genetic etiology of CAD [25].

2.2 Low Dietary Fiber of Polished Foods is Key Factors that Cause Heart Disease

The loss of fibre from brown rice to polished rice of China and world in 2010 is 3,964,955 tons and 13,926,233 tons, however whole grains to refined flour of wheat is 13,510,200 tons and 78,476,640 tons, respectively [26]. Dietary fiber is associated with lowering the risk of CHD, diabetes, cancers and other chronic diseases [27]. High carbohydrate intake of 70% from polished rice and 17% from refined products of wheat is associated with increased the risk of CHD in Chinese adults [28], and has significant correlation between wheat flour with high GI and CHD in China. The replacement of rice + noodles every day was associated with higher risk (26.11%), but rice + vegetables (-23.81%), fruit (-11.94%), or whole bread (-19.46%) every day was associated with lower risk of IHD death [29]. A large consumption of dietary fiber from cereals or vegetables is associated with lower risk of fatal IHD [30].

2.3 Obesity is A Key Risk Factor of Cause Heart Disease

Hypertension, diabetes, some cancer and heart disease are associated with obesity [26,31-34]. Obesity increases the risk of chronic diseases, such as heart disease, diabetes II, insomnia, some cancer and osteoarthritis [35]. More than 200 million hypertension in China was second major cause of heart failure [36]. The risk ratio each 5 kg/m^2 higher body-mass index was 1.27 for CHD and 1.18 for stroke after adjustment for confounders [37]. Adiponectin deficiency can increase fat, induced obesity, metabolic disorder, cardiac aberrance through decreased myocardial autophagy [38]. Hypertension is an important risk factor for CAD, its four loci *(SH2B3, GOSR2, YP17A1-NT5C2, CUCY1A3-GUCY1B3)* displayed study-wide significant association with CAD [39].

2.4 High Sodium and Phosphorus Intakes are the Major Factor of Cause Heart Disease

Hypertension and heart disease are frequently associated with Na. Fibroblast growth factor-23 can regulate renal phosphate reabsorption and vitamin D synthesis, affect renal Na retention, hypertension and heart hypertrophy [40]. Both low Na (<115 mmol) and high Na intakes (>215 mmol) consistent with a U-shaped association between Na intake and health, which is associated with increased mortality [41]. A higher urinary Na / K excretion ratio is remarkably associated with ventricular arrhythmias among well-controlled blood pressure [42]. QTLs with pleiotropic effects for *qK1/qMg1/ qCa1* region between markers Bmag0211 and GBMS0014 on chromosome 1H in barley was shown to have large additive effects in grains [43]. The variability of serum P is associated with coronary artery calcification and keeping serum P stable may decrease morbidity and mortality in patients [44]. Higher dietary P intake was associated with left ventricular mass [45]. Low urinary Mg excretion was associated with risk of IHD [46]. Women consumed more than 1400 mg of Ca daily had a higher risk of death from IHD but not stroke [47]. High serum transferrin saturation concentration was contrary inversely associated with CHD [48].

3 Natural Functional Food for Heart Disease Prevention

Heart disease is one of the most important diseases threatening human health. Prevention and control of heart disease rely on adopting a balanced diet, including mineral elements, protein, fats, and fiber from whole grains, vegetables and fruits.

3.1 Functional Food with High Dietary Fiber from Brown Rice or Whole Wheat Flour are the Key to Heart Diseases Prevention in Human Being.

In 2013, global rice yield and polished rice were 712.7 million tons and 477.5 million tons, but Chinese rice yield and polished rice were 202.75 million tons and 136.4 million tons, respectively. The loss of dietary fiber from brown rice to polished rice of China and throughout the world in 2013 were 4,056,746 tons and 13,981,979 tons, however whole grains to refined flour of wheat is 14,297,316 tons and 81,880,729 tons, respectively (See Table 1,2 and 3). Increasing consumption of cereal fiber when myocardial infarction was remarkably associated with lower cardiovascular mortality [49]. The loss of potassium from brown rice to polished rice of China and throughout the world in 2013 was 269,763 tons and 929,765 tons, however whole grains to refined flour of wheat is 835,128 tons and 4,782,778 tons, respectively (See Table 1, 2 and 3). Fibroblasts play a key role in cardiac function / dysfunction, regulating fibroblast K^+ channels can prevent fibrosis and atrial fibrillation [50]. The loss of magnesium from brown rice to polished rice of China and the entire world in 2013 were 140,972 tons and 485,875 tons, however whole grains to refined flour of wheat is 247,010 tons and 1,414,626 tons, respectively (See Table 1,2 and 3). Compared with lower consumption, population in the highest Mg intake had a 34% reduction risk in mortality of cardiovascular disease [51], especially heart disease. The consumption of whole grains (wheat or rice) account for 1% but refined flour or polished rice account for 96% in China [26], however the consumption of whole grains (wheat or rice) account for 10% but refined flour or polished rice account for 85% in the world. Barley grass powder with higher GABA, Ca and K is the most ideal functional food promoting sleep [52]. Therefore, although whole grain (brown rice and wheat) with fiber, potassium and magnesium as well as B vitamins, make them for the useful food to foster heart health, but the replacement of major refined products (rice, wheat) with whole grains (brown rice, whole wheat) is very difficult.

Table 1: The Elements Concentration Loss from Brown Rice to Polished Rice Based on 202.75 Million Tons of Rice Yield in Ch ina and 712.7 Million Tons of Rice Yield of The World in 2013 [25] (http://www.fas.usda.gov/)

Elements	Elements content loss (%)	China (tons)	Globe (tons)
Strontium (Sr)	99.3	711,2	24,512
Chromium (Cr)	99.2	18,364	63,294
Magnesium (Mg)	61.8	140,972	485,875
Zinc (Zn)	61.0	7,007	24,149
Molybdenum (Mo)	60.6	350	1,206
Phosphorus (P)	57.9	399,036	1,375,320
Potassium (K)	55.9	269,763	929,765
Manganese (Mn)	55.0	2,982	10,278
Tin (Sn)	54.3	234	807
Nickel (Ni)	42.9	167	574
Iron (Fe)	31.7	1,113	3,838
Calcium (Ca)	31.2	9,168	31,599
Sodium (Na)	10.8	384	1,322
Sulphur (S)	9.1	17,102	58,944

Table 2: The Functional Ingredients Loss from Brown Rice to Polished Rice Based on 202.75 Million Tons of Chinese Rice Yield and 712.7 Million Tons of Rice Yield OF The World in 2013[25] (http://www.fas.usda.gov/)

Functional ingredients	Ingredients loss (%)	China (tons)	Globe (tons)
Dietary fibre	57.0	4,056,746	13,981,979
Total flavones	79.2	162,325	559,469
Total alkaloids	40.6	55,917	192,724
GABA	78.8	20,403	70,323
Pantothenic acid	32.0	972	3,350
Pyridoxine	68.0	704	2,426
Thiamine	83.0	502	1,731
Riboflavin	84.0	159	547

Table 3: The Nutrition Loss from Whole Grains to Refined Flour of Wheat Based on 121.7 Million Tons of Wheat Yield in China and 697 Million Tons of Global Wheat Yield in 2013[25] (http://www.fas.usda.gov/)

Nutrition components	Components loss (%)	China (tons)	Globe (tons)
Phosphorus (P)	91	931,646	5,335,536
Potassium (K)	77	835,128	4,782,778
Magnesium (Mg)	85	247,010	1,414626
Calcium (Ca)	60	28,452	162,944
Zinc (Zn)	98	14,644	83,866
Manganese (Mn)	86	13,909	79,657
Iron (Fe)	75	5,708	32,690
Dietary fiber	89	14,297,316	81,880,729
Niacin	81	6,710	3,8428
Thiamine (VB1)	77	1,798	10,290
Pyridoxine (VB6)	72	1,138	5,517
Riboflavin (VB2)	67	407	2,314
Pantothenic acid	50	31	118

3.2 Functional Foods with Low Glycemic Index from Barley and its Grass Powder or Oat are the Key to Heart Diseases Prevention in Human Being

Women who eat foods with high glycemic index (GI), such as wheat flour, may be at greater risk for heart disease [53]. A QTL *(CslF6)* explaining 31% of the genet of β-glucan was located on 7HL, and naked barley for functional foods had significantly lower GI than oats [54]. The consumption of barley β-glucan alters microorganisms associated with a reduced risk of cardiovascular disease [55]. The wholegrain oat can be prebiotics and have low GI (40) [56]. Mean dietary GI was 56.2 approximately normal, however mean GI of major foods were 87 for polished rice and 86 for wheat flour, but 55 brown rice and 41 whole wheat flour as well as 25 pearl barley [57]. Tsangpa for barley products is taken as a baiyao, based on lower rate of heart disease and colon cancers occur in Tibet than that of expected [58]. The potential magnitude of health-related cost savings by health claims for soluble fiber of barley products and CHD [59]. β-glucan of barley reduces the risk of CHD; arabinoxylan and resistant starch can ameliorate glycemic control [60]. Barley spaghetti for flavan-3-ols and soluble fiber as well as β-glucan enrichment reached USFDA requirements, which could allow these products for health claims “good source of dietary fiber”, “may reduce the risk of heart disease” [61]. USA and UK allowed health claim that oats β-glucan can reduces risk of heart disease and plasma cholesterol levels [62]. The intake of five bioactive substances controlled in whole grains (wheat, barley, oat) and their fractions is associated with a decreased risk of CHD and diabetes, which includes β-glucans, arabinoxylans, alkylresorcinols, tocols and phytosterols [3]. Whole grains (barley, oats) with high in viscous fiber can reduce the risk for CHD, serum low-density lipoprotein cholesterol, and blood pressure; meanwhile improving glucose and insulin responses [63]. Barley grass powder every day resulted in heart disease prevention, improved sleeping, lustihood, regulated blood sugar and pressure, enhanced immunity and liver function, detoxification acne skin, prevention of constipation, repair memory, improved gastrointestinal function, reduced gout and hyperuricemia disease, bone injury recovery, alleviated nerve dermatitis, anti-cancer, anti-aging, losing weight and reducing blood fat, anti-inflammatory and so on [26,64]. The reason is that the product is rich in nutritional and functional components, compared with brown rice, its total flavonoids and alkaloids were 2.1 times and 10.7 times, respectively; especially γ-aminobutyric acid for 37.8 times and potassium for 13 times, calcium for 55 times, iron for 30.2 times of brown rice [64].

3.3 Functional Foods with High Bioactive Components from Vegetables or Fruits are the Key to Heart Diseases Prevention in Human Being

The increased consumption of fruits and vegetables was deemed to be protecting humans against cancer, diabetes, heart and brain vascular diseases, which due to the effect of α-tocopherol, ascorbic acid, β-carotene, polyphenols and anthocyanins [65]. The dietary capsaicin can control cardiac hypertrophy and fibrosis in pressure overburden mice [66]. Ginger is a good functional food for treatment of hypertension and heart disease [2]. Garlic can protect the heart against myocardial infarction, hypertrophy, arrhythmia, cardiotoxicity and so on, based on H_2S and NO (nitric oxide) in cardiomyocytes and endothelial cells [67]. Some results support that the health effects of canola oil on CHD, insulin sensitivity, inflammation, and cancer cell growth [68]. Consumption of white vegetables with high potassium related to decreased risk of CHD and stroke [69]. Mortality from IHD and circulatory as well as cerebrovascular diseases was significantly lower in vegetarians with plenty of vegetables and fruits than in omnivorous populations [70]. An increased consumption of whole grain foods and fruits as well as vegetable leads to a reduced risk of multimorbidity [71]. Walnut area of Yunnan province in China reached 2646667 hectares, output and its value in 2013 have reached 680000 tons and 16 billion yuan; Consumption of walnut with high polyphenols related to decreased risk of CHD and other chronic diseases [72]. Olive and olive oil can prevent CHD and certain cancers based on their abundant monosaturated fatty acids and phenolic compounds [73]. Olive fruit contains four hydrophilic and one lipophilic phenolic compounds with cardiotonic, anti-hypertensive, antioxidant, anti-carcinogenic, anti-inflammatory, anti-microbial, laxative, anti-dyslipidemic, and antiplatelet [74]. Oral *Cordyceps sinensis* remarkable attenuates the heart injuries in chronic kidney disease rats [75]. The risk of myocardial infarction and cardiovascular disease is associated with chromosome *9p21* SNPs, which appears to be altered by the sufficient raw vegetables and fruits [76]. The magnanimous fruit and vegetable intake can reduce the risk of CHD by 17% [77]. *Capsella bursapastoris* is used as reduction of blood pressure, inhibition of inflammation, anti-cancer and diuretics [78], based on its bioactive components such as glycosides of quercetin, chrysoeriol, kaempferol, and isorhamnetin [79]. The proanthocyanidin in grape seed can offer cardiac protection against IHD [80].

3.4 Functional Foods with High Bioactive Components from Vegetables or Fruits are the Key to Heart Diseases Prevention in Human Being

Potassium such as in banana is important in physiological homeostatic control of cardiac function [81]. Selenium (Se) is a biomarker of CHD, meanwhile Se supplementation is very important to patients undergoing cardiac surgery [82]. Increasing the contractility of heart muscle cells by boosting intracellular Ca handling is an effective therapy for heart failure, however miR-25 overexpression causes declining cardiac function [16]. Troponin is integral to contraction of skeletal and cardiac muscle, and polymer paper to detect heart disease [83]. Supplementary Ca intake related to risk of CHD [84]. Dietary Mg intake was inversely related with reduced risk of CHD and IHD [46,85]. Decreasing dietary salt intake reduces CHD deaths in the four counties (Tunisia, Syria, Palestine, and Turkey), however a comprehensive strategy can save money ($13.28 billion) and lives (419250) [86]. Garlic and onion may be the organopolysulfides and quercetin mechanism in the treatment of chronic diseases [87].

3.5 Mushrooms is One of the Most Important Ways of Heart Diseases Prevention

Ganoderma tsugae possess the cardioprotective activity and ganoderic acids against cardiac insults [88]. *Agaricus brasiliensis* polysaccharide includes glucose (78.38%), arabinose (10.46%) and mannose (8.51%), which was propitious to some cardiovascular diseases [89] and used to treat heart diseases as well as prevent cancer [90]. *Lentinus edodes* is the treatment of chronic diseases involving heart disease, cancer, hypertension, diabetes, hyperlipidemia, hepatitis and so on, based on bioactive components such as lentinan, eritadenine, and lectins [91]. Cordyceps can treat heart disease, arrhythmias, hyperglycaemia, hyperlipidaemia, hyposexuality, and so on, based on three bioactive components (3'-deoxyadenosine, cordycepic acid and polysaccharides) [92]. Some mushrooms with abundant potassium without sodium can suit for functional foods of hypertension and patients with heart disease [90].

4 Discussion and Future research Questions

4.1 Large-Scale Chronic Disease Differences for Rice Versus Wheat Agriculture

Asian countries have more cases of stroke and diabetes than that of CHD, whereas the opposite is true in Western countries, based on 10 Asian countries and the United Kingdom as well as the United States [93, 94]. Although similar yields between rice and wheat in the world in 2013, but the loss of dietary fiber from whole wheat to wheat flour is 5.9 times than that of brown rice to white rice, however the loss of potassium and magnesium from whole wheat to wheat flour are 5.1 times and 2.9 times than that of brown rice to white rice, respectively (See Table 1, 2, 3). Higher potassium and magnesium as well as dietary fiber are associated with decreased risk of CHD; polished rice is major food of Asian populations, but wheat flour is major food of Western populations. Higher consumption of polished rice is related to a remarkable increased risk of diabetes II for Asian populations [95], but functional rice with high resistant starch is an important ways for diabetes preventing in China and Asia [96,97]. The Near East Fertile Crescent is the center of origin, diversity, and domestication for barley, whereas Tibet in China is one of domestic centers of cultivated barley [98]. This is one reason that Western countries have more CHD than that of stroke and diabetes for Asian countries, based on the loss of K and Mg as well as dietary fiber of major food from whole wheat to wheat flour, whereas another reason that Asian countries have more stroke and diabetes than that of CHD for Western countries, based on Asian diabetes outbreak is rooted in brown rice and barley with low GI (25) for major food of the ancient humans into polished rice and white flour with high GI (87) as staple food of modern humans [64]. In addition, farming rice makes lots of cultures, whereas farming wheat more independent [99]. In the future research, we will discover that coevolution between heart disease and food structure of human being for Asia and Western countries.

4.2 Chronic Disease especially Heart Disease for Modern Humans Associated with Food Structure of Ancient Humans

The dietary flexibility of early hominins to include foods consumption from grasses, sedges, and succulents in tropical savannas and deserts represents a remarkable ecological and behavioral distinction from extant great apes as well as the last common ancestor [100]. Utilization of functional foods for dietary of chronic disease preventing in human being includes whole grains (brown rice, wheat flour, barley, buckwheat, oats, and others) or functional rice with high resistant starch food, functional vegetables (bitter melon, garlic, onions, broccoli, cabbage, and others), functional fruits (malnut, blueberries, strawberries, watermelon, olive oil, and others), mushrooms, green tea, coffee, and barley grass powder [26,32,33]. We can guess that chronic disease especially heart disease of human being was associated with the five evolutionary stages of the major dietary, which was as follows: I-fruits or vegetables, II-grass or *Cyperaceou*, III-cereals (rice, wheat, millet, beans, barley and corn), IV-polished rice or wheat flour, V-polished rice or wheat flour+barley grass powder. In the future research, we will discover that correlation evolution between heart disease and five food structure of human being.

4.3 Origin of Functional Crop Evolution is Closely Related to Chronic Diseases for Dietary of Human Being

Functional crop (brown rice, vegetables, fruits, and mushrooms) for preventing heart disease is not only similar to cancer [33] and hypertension [32] as well as diabetes prevention [26] in China and in the world, but also further support that Southwest China (especially Yunnan Province) is one common sphere that the coevolution of functional crop is closely related to the chronic diseases of human.

The Near East Fertile Crescent is not only one of the earliest domestication of crop in the world, but also the center of origin and diversity of wild wheat and barley [26], the cradle for human civilization. Barley grass powder is a remarkable decrease in fasting blood sugar, total cholesterol, low-density lipoprotein cholesterol, but a remarkable increase in high-density lipoprotein cholesterol levels [101]. The diet of *Theropithecus* for a common large-bodied primate was making up of grasses or sedges, which occurred with hominins in East and South Africa [102]. The human's evolution of an enlarging brain, a contracting large intestine, and prolonging small intestine necessitated a require for nutritionally dense foods (Armelagos 2014). In future research, it is necessary to discover the origin and evolution of functional crop associated with chronic diseases of human evolution.

Acknowledgments: This research was supported by China Agriculture Research System (CARS-05), the National Natural Science Foundation of China (No.31260326), the Science and Technology to Benefit the People (2014RA060) and the Exploitue of Emphases New Production (2014BD001) from Yunnan Provincial Scientific and Technology Department.

References

[1] Majed B., Arveiler D., Bingham A., Ferrieres J., Ruidavets J. B., et al.,Depressive symptoms, a time -dependent risk factor for coronary heart disease and stroke in middle-aged men:The PRIME Study[J], Stroke, 2012, 43(7), 1761-1767.

[2] Akinyemi, A.J., Ademiluyi A.O., Oboh, G., Inhibition of angiotensin-1-converting enzyme activity by two varieties of ginger (*Zingiber officinale*) in rats fed a high cholesterol diet[J], J. Med. Food, 2014, 17(3), 317-323.

[3] Bartłomiej S., Justyna R.K., Ewa N., Bioactive compounds in cereal grains - occurrence, structure, technological significance and nutritional benefits - a review[J]. Food Sci. Technol. Int.,2012,18 (6), 559-568.

[4] Franklin B.A., Durstine J.L., Roberts C.K.U., Barnard R.J., Impact of diet and exercise on lipid management in the modern era[J], Best Pract. Res. Clin Endocrinol. Metab., 2014, 28(3), 405 - 421.

[5] Dave V.P., Kaul D., Heart and Brain: A neutro-genomic link[J], J. biolog. Chem., 2013,4(2), 6-17.

[6] Johnson W.H.J.,Moller J.H., Pediatric Cardiology/The Essential Pocket Guide: A healthy lifestyle and preventing heart disease in children[C], Chichester:John Wiley & Sons, Ltd, UK, 2014,doi:10.1002/978111 8503379.ch12.

[7] Dalen J.E., Devries, S., Diets to prevent coronary heart disease 1957-2013: what have we learned?[J], American J. Med. 2014, 127(5), 364-369.

[8] Garcia-Rios A., Delgado-Lista J., Alcala-Diaz J.F., Lopez-Miranda J., Perez-Martinez P., Nutraceuticals and coronary heart disease[J], Curr. Opin. Cardiol.,2013, 28 (4), 475-482.

[9] Szakály Z., Szente V., Kövér G., Polereczki Z., Szigeti O.,The influence of lifestyle on health behavior and preference for functional foods[J], Appetite,2012,58(1), 406-413.

[10] Huang W.Y., Davidg S.T., Wu, J., Bioactive natural constituents from food sources-potential use in hypertension prevention and treatment[J], Crit. Rev. Food Sci. Nutr., 2013, 53(6), 615-630.

[11] Vella M.N., Stratton L.M., Sheeshka J.,Duncan A.M., Exploration of functional food consumption in older adults in relation to food matrices, bioactive ingredients, and health[J], J. Nutr. Gerontol. Geriatr., 2013, 32(2), 122-144.

[12] Hisamatsu T., Miura K., Ohkubo T., Yamamoto T., Fujiyoshi A.,et al.,High long-chain n-3 fatty acid intake attenuates the effect of high resting heart rate on cardiovascular mortality risk: A 24-year follow-up of Japanese general population[J],J. Cardiol., 2014, 64(3), 218-224.

[13] Najafi M., Sheikhvatan M.. Gender differences in coronary artery disease: correlational study on dietary pattern and known cardiovascular risk factors[J], Int. Cardiovasc. Res. J., 2013, 7(4), 124-129.
[14] Song C.,Chang Z., Magnusson P.K.E., Ingelsson E.,Pedersen N.L., Genetic factors may play a prominent role in the development of coronary heart disease dependent on important environmental factors[J], J. Intern. Med., 2014, 275(6), 631-639.
[15] Harrison C., Cardiovascular disease: Inhibiting microRNA-34 benefits heart disease[J], Nat. Rev. Drug Discov., 2012, 11(908), 1474-1776.
[16] Wahlquist C., Jeong D., Rojas-Muñoz A., Kho C., Lee A., Mitsuyama S.,et al., Inhibition of miR-25 improves cardiac contractility in the failing heart[J], Nature, 2014, 508(7497), 531-535.
[17] Ghosh T.K., Granados-Riveron J.T., Buxton S., Setchfield K., Loughna S., Brook J.D.,Studies of genes involved in congenital heart disease[J], J. Cardiovasc. Develop. Dis., 2014, 1(1), 134 -145.
[18] Miller C.L., Haas U., Diaz R., Leeper N.J., Kundu R.K., Patlolla B.,et al., Coronary heart disease-associated variation in TCF21 disrupts a miR-224 binding site and miRNA-mediated regulation[J], PLoS Genet., 2014, 10(3), e1004263.
[19] Zhang M.D., Gu W., Qiao S.B., Zhu E.J., Zhao Q.M., Lv S.Z., Apolipoprotein e gene polymorphism and risk for coronary heart disease in the chinese population: a meta-analysis of 61 studies including 6634 cases and 6393 controls[J], PLoS One, 2014, 9(4), e95463.
[20] Cao Y., Wang J., Wei C., Hou Z., Li Y., Zou H.,et al., Genetic variations of NKX2-5 in sporadic atrial septal defect and ventricular septal defect in Chinese Yunnan population[J], Gene,2016, 575 (1), 29-33.
[21] Zheng R.L., Zhang H., Jiang W.L.,Tumor necrosis factor-alpha 308G>A polymorphism and risk of rheumatic heart disease: a meta-analysis[J], Scientific reports, 2014. 4,4731.
[22] Zhao J.Y., Yang X.Y., Shi K.H., Sun S.N., Hou J., Ye Z.Z.,et al., A functional variant in the cystathionine β-synthase gene promoter significantly reduces congenital heart disease susceptibility in a Han Chinese population[J], Cell Res., 2013, 23(2), 242-253.
[23] Abunimer A., Smith K., Wu T.J., Lam P., Simonyan V., Mazumder R., Single-nucleotide variations in cardiac arrhythmias: prospects for genomics and proteomics based biomarker discovery and diagnostics[J], Genes (Basel),2014, 5(2), 254-269.
[24] Lu X., Wang L., Chen S., He L., Yang X., Shi Y.,et al., Genome-wide association study in Han Chinese identifies four new susceptibility loci for coronary artery disease[J], Nat. Genet., 2012, 44(8), 890-894.
[25] The CARDIoGRAMplusC4D Consortium.,Large-scale association analysis identifies new risk loci for coronary artery disease[J], Naure, Genetics, 2013, 45(1), 25-33.
[26] Zeng Y.W., Pu X.Y., Du J., Yang S.M., Yang T.,Jia P., Use of functional foods for diabetes prevention in China[J], Afr. J. Pharm. Pharmacol. 2012, 6(35), 2570-2579.
[27] Park Y., Subar A.F., Hollenbeck A., Schatzkin A., Dietary fiber intake and mortality in the NIH-AARP diet and health study[J], Arch. Intern. Med., 2011, 171(12), 1061-1068.
[28] Yu D., Shu X.O., Li H., Xiang Y.B., Yang G., Gao Y.T., et al., Dietary carbohydrates, refined grains, glycemic load, and risk of coronary heart disease in Chinese adults[J], Am. J. Epidemiol., 178(10), 1542-1549.
[29] Rebello S.A., Koh H., Chen C., Naidoo N., Odegaard A.O., Koh W.P., et al., Amount, type, and sources of carbohydrates in relation to ischemic heart disease mortality in a Chinese population: a prospective cohort study[J], Am. J. Clin. Nutr., 2014, 100(1), 53-64.
[30] Crowe F. L., Key T. J., Appleby P. N., Overvad K., Schmidt E. B., Egeberg R.,et al., Dietary fibre intake and ischaemic heart disease mortality:the European rospective investigation into cancer and nutrition -heart study[J], Eur. J. Clin. Nutr.,2012, 66(8), 950 - 956.
[31] Tonelli A. R., A coat connects obesity with heart disease[J], Sci. Translat. Med., 2013, 5(188), 188ec95.

[32] Zeng Y.W., Du J., Pu X.Y., Yang S.M., Yang T., et al., Strategies of functional food for hypertension prevention in China[J], J. Med. Plant Res., 2011, 5(24), 5671-5676.
[33] Zeng Y.W., Yang J.Z., Pu X.Y., Du J., Yang T., Yang S.M.,et al., Strategies of functional food for cancer prevention in human being[J], Asian Pac. J. cancer prev., 2013, 14(3), 1585-1592.
[34] Zeng Y.W., Du J., Pu X.Y., Yang J.Z., Yang T., Yang S.M., Yang X.M.. Coevolution between Cancer Activities and Food Structure of Human Being from Southwest China[J]. BioMed Res. Int.,2015, 2015 ID 497934, 1-10.
[35] Haslam D.W.,James W.P., Obesity, Lancet, 2005, 366(9492), 1197-1209.
[36] Su G.H., Cao H., Xu S.D., Lu Y.X., Shuai X.X., Sun Y.F.,et al., Left Atrial enlargement in the early stage of hypertensive heart disease: a common but ignored condition[J],J. Clin. Hypertens., 2014, 16(3), 192-197.
[37] The Global Burden of Metabolic Risk Factors for Chronic Diseases Collaboration., Metabolic mediators of the effects of body-mass index, overweight, and obesity on coronary heart disease and stroke: a pooled analysis of 97 prospective cohorts with 1.8 million participants[J], *The* Lancet, 2014, 383(9921), 970-983.
[38] Guo R., Zhang Y., Turdi S., Ren J., Adiponectin knockout accentuates high fat diet-induced obesity and cardiac dysfunction: role of autophagy[J], Biochim. Biophys. Acta, 2013, 1832(8), 1136 1148.
[39] Lieb W., Jansen H., Loley C., Pencina M.J., Nelson C.P., Newton-Cheh C.,et al., Genetic predisposition to higher blood pressure increases coronary artery disease risk[J], Hypertension, 2013, 61(5), 995-1001.
[40] Andrukhova O., Slavic S., Smorodchenko A., Zeitz U., Shalhoub V.,et al., FGF23 regulates renal sodium handling and blood pressure[J]. EMBO Mol. Med., 2014, 6(5), 569-703.
[41] Graudal N., Jürgens G., Baslund B., Alderman M.H., Compared with usual sodium intake, low- and excessive-sodium diets are associated with increased mortality: a meta-analysis[J], Am. J. Hypertens., 2014, 27(9), 1129–1137.
[42] Marketou M.E., Zacharis E.A., Parthenakis F., Kochiadakis G.E., Maragkoudakis S., Chlouverakis G., Association of sodium and potassium intake withventricular arrhythmic burden in patients with essential hypertension[J],J. Am. Soc. Hypertens., 2013, 7(4), 276-282.
[43] Zeng Y.W., Du J., Yang X.M, Pu X.Y., Wang L.X., Yang J.Z., et al. Identification of quantitative trait loci for mineral elements in grains and grass powder of barley[J]. Genet. Mol. Res., 2016, 15(4): gmr15049103.
[44] Wang, M.J.; Li, H.M.; You, L.; Yu, X.L.; Zhang, M.; Zhu, R.J.; Hao, M.; Zhang, Z.J. & Chen, J.2014. Association of serum phosphorus variability with coronary artery calcification among hemodialysis patients. *PLOS one* 9(4): e93360.
[45] Yamamoto K.T., Robinson-Cohen C., de Oliveira M.C., Kostina A., Nettleton J.A., Ix J.H., et al., Dietary phosphorus is associated with greater left ventricular mass[J], Kidney Int., 2013, 83(4), 707-714.
[46] Joosten M.M., Gansevoort R.T., Mukamal K.J., van der Harst P., Geleijnse J.M., Feskens E.J., et al., Urinary and plasma magnesium and risk of ischemic heart disease[J], Am. J. Clin. Nutr.,2013, 97(6), 1299-1306.
[47] Kuehn B.M., High calcium intake linked to heart disease, death[J], J. Am. Med. Assoc., 2013, 309(10), 972.
[48] Hunnicutt J., He K.,Xun P., Dietary iron intake and body iron stores are associated with risk of coronary heart disease in a meta-analysis of prospective cohort studies[J], J. Nutr., 2014, 144(3), 359-366.
[49] Li S., Flint A., Pai J. K., Forman J.P., Hu F.B., Willett W.C.,et al., Dietary fiber intake and mortality among survivors of myocardial infarction: prospective cohort study[J], Br. Med. J., 2014,348, g2659.

[50] Wu C.T., Qi X.Y., Huang H., Naud P., Dawson K., Yeh Y.H., et al., Disease and region-related cardiac fibroblast potassium current variations and potential functional significance. Cardiovasc. Res., 2014, 102(3), 487-496.
[51] Guasch-Ferré M., Bulló M., Estruch R., Corella D., Martínez-González M.A.,et al.,Dietary magnesium intake is inversely associated with mortality in adults at high cardiovascular disease risk[J],. J.Nutr., 2013, 144(1), 55-60.
[52] Zeng Y.W.,Yang J,Z., Du J.,Pu X.Y.,Yang X.M.,Yang S.M., et al., Strategies of functional foods promote sleep in human being[J]. Curr. Signal Transd. T., 2014, 9(3): 148-155.
[53] Sieri S., Krogh V., Berrino F., Evangelista A., Agnoli C., Brighenti F., et al., Dietary glycemic load and index and risk of coronary heart disease in a large Italian cohort: The EPICOR study[J], Arch. Intern. Med. 2010. 170 (7), 640-647.
[54] Steele K., Dickin E., Keerio M.D., Samad S., Kambona C., Brook R.,et al., Breeding low-glycemic index barley for functional food[J], Field Crops Res., 2013, 154, 31-39.
[55] Wang Y., Ames N.P., Tun H.M., Tosh S.M., Jones P.J., Khafipour E., High molecular weight barley β-glucan alters gut microbiota toward reduced cardiovascular disease risk[J], Front. microbiol., 2016,.7, 129.
[56] Connolly M.L.,Tuohy K.M., Lovegrove,J.A.,Wholegrain oat-based cereals have prebiotic potential and low glycaemic index[J], Br. J. Nutr.,2012, 108(12), 2198-2206.
[57] Lin C.S., Kimokoti R.W., Brown L.S., Kaye E.A., Nunn M.E., Millen B.E., Methodology for adding glycemic index to the National Health and Nutrition Examination Survey nutrient database[J], J. Acad. Nutr. Diet., 2012, 112(11), 1843-1851.
[58] Nyima T., Tang Y.W., Zeng X.Q., In: G. P. Zhang, C.D.Li, & X.Liu, (Eds.). Advance in Barley Sciences: Food preparation from hulless barley in Tibet. Proceedings of 11th International Barley Genetics Symposium[C], Hangzhou, China: Zhejiang University Press, 2012, pp. 92-95.
[59] Malla S., Hobbs J.E.,Sogah E.K., Functional foods, health benefits and health claims[J], Athens J. Healt, 2014, 1(1), 37-46.
[60] Bernstein A.M., Titgemeier B., Kirkpatrick K., Golubic M., Roizen M.F., 2013.Major cereal grain fibers and psyllium in relation to cardiovascular health[J],Nutrients, 2013,5(5), 1471 - 1487.
[61] Verardo V., Gómez-Caravaca A.M., Messia M.C., Marconi E., Caboni M.F., Development of functional spaghetti enriched in bioactive compounds using barley coarse fraction obtained by air classification[J], J. Agri. Food Chem., 2011, 59(17), 9127-9134.
[62] Othman R.A., Moghadasian, M.H., Jones P.J., Cholesterol-lowering effects of oat β-glucan[J], Nutr. Rev., 2011, 69(6), 299-309.
[63] Harris K.A.,Kris-Etherton P,M.,Effects of whole grains on coronary heart disease risk[J], Curr. Atheroscler. Rep.,2010, 12(6), 368-376.
[64] Zeng Y.W., Pu X.Y., Zhang J., Guo G.G., Du J., Yang T., et al., Synthetic research and utilization on industrial development of barley in Southwestern China[J], J. Agri. Sci. Technol., 2013, 15(3), 48-56.
[65] Harasym J., Oledzki, R., Effect of fruit and vegetable antioxidants on total antioxidant capacity of blood plasma[J], Nutrition, 2014, 30(5), 511-517.
[66] Wang Q., Ma S.T., Li D., Zhang Y., Tang B., Qiu C.M.,et al., Dietary capsaicin ameliorates pressure overload–induced cardiac hypertrophy and pibrosis through the transient receptor potential vanilloid type 1[J], Am. J.Hypertens., 2014, 27(12), 1521-1529.
[67] Khatua T.N., Adela R.,Banerjee S.K.,2013.Garlic and cardioprotection: insights into the molecular mechanisms. Can. J. Biochem. Physiol., 91(6), 448-458.
[68] Lin L., Allemekinders H., Dansby A., Campbell L., Durance-Tod S., Berger A., Evidence of health benefits of canola oil[J], Nutr. Rev. 2013, 71(6), 370-385.
[69] Weaver C.M., Potassium and health[J], Adv. Nutr., 2013, 4(3), 368S-377S.
[70] Li D., Effect of the vegetarian diet on non-communicable diseases[J], J. Sci. Food Agric., 2014, 94(2), 169-173.

[71] Ruel G., Shi Z., Zhen S., Zuo H., Kröger E, Sirois C., et al., Association between nutrition and the evolution of multimorbidity: The importance of fruits and vegetables and whole grain products[J], Clin. Nutr. 2014,33(3),513-520.
[72] Regueiro J., Sánchez-González C., Vallverdú-Queralt A., Simal-Gándara J., Lamuela - Raventós R., Izquierdo-Pulido M., Comprehensive identification of walnut polyphenols by liquid chromatography coupled to linear ion trap-Orbitrap mass spectrometry[J], Food Chem., 2014. 152,340-348.
[73] Uylaşer V.,Yildiz G., The historical development and nutritional importance of olive and olive oil constituted an important part of the Mediterranean diet[J], Crit. Rev. Food Sci. Nutr., 2014,54(8),1092-1101.
[74] Ghanbari R., Anwar F., Alkharfy K.M., Gilani A.H.,Saari, N., Valuable nutrients and functional bioactives in different parts of olive (*Olea europaea* L.)-A review[J], Int. J. Mol. Sci., 2012,13(3), 3291-3340.
[75] Liu X., Zhong F.,Tang X., Lian F.L., Zhou Q., Guo S.M., et al.,*Cordyceps sinensis* protects against liver and heart injuries in a rat model of chronic kidney disease: a metabolomic analysis[J], Acta Pharmacol. Sin., 35(5), 697 - 706.
[76] Do R., Xie C.C., Zhang X.H., Männistö S., Harald K., Islam S.,et al.,The effect of chromosome 9p21 variants on cardiovascular disease may be modified by dietary intake: evidence from a case/control and a prospective study[J], PLoS Med.,2011, 8(10),e1001106.
[77] Bhupathiraju S.N., Wedick N.M., Pan A., Manson J.E., Rexrode K.M., Willett W.C., et al., Quantity and variety in fruit and vegetable intake and risk of coronary heart disease[J], American J. Clin. Nutr., 2013, 98(6), 1514-1523.
[78] Lee K.E., Shin J.A., Hong I.S., Cho, N.P., Cho, S.D.,Effect of methanol extracts of *Cnidium officinale Makino* and *Capsella bursa-pastoris* on the apoptosis of HSC-2 human oral cancer cells[J], Exp. Ther. Med.,2013, 5(3), 789-792.
[79] Kubínová R., Spačková V., Svajdlenka E., Lučivjanská K., Antioxidant activity of extracts and HPLC analysis of flavonoids from *Capella bursa-pastoris* (L.) Medik[J], Ceska Slov. Farm., 2013,62(4),174-176.
[80] Wang C.Z., Mehendale S.R., Calway T., Yuan C.S., Botanical flavonoids on coronary heart disease[J], Am. J. Chin. Med., 2011,39(4), 661-671.
[81] Khan E., Spiers C.,Khan M., The heart and potassium: a banana republic[J], Acute Cardiovasc. Care, 2013, 15(1),17-24.
[82] Benstoem C., Goetzenich A., Kraemer S., Borosch S., Manzanares W., Hardy G., et al., 2015. Seleniumand its supplementation in cardiovascular disease--what do we know?[J], Nutrients, 2015,7 (5), 3094-3118.
[83] Jagadeesan K.K., Kumar S., Sumana G., Application of conducting paper for selective detection of troponin[J], Electrochem. Commun. 2012,20, 71-74.
[84] Umesawa M.,Iso, H., Associations between calcium intake and risk of cardiovascular disease[J], Clin. Calcium, 2013,23(4), 583-588.
[85] Chiuve S.E., Sun Q., Curhan G.C., Taylor E.N., Spiegelman D.,et al., Dietary and plasma magnesium and risk of coronary heart disease among women[J], J. Am. Heart Assoc.,2013,2(2),e000114.
[86] Mason H., Shoaibi A., Ghandour R., O'Flaherty M., Capewell S., Khatib R.,et al., A cost effectiveness analysis of salt reduction policies to reduce coronary heart disease in four Eastern Mediterranean countries[J]. PLoS One, 2014,9(1), e84445.
[87] Zeng Y.W., Li Y.P., Yang J.Z., Pu X.Y., Du J., Yang X.M., et al., Therapeutic role of functional components in Alliums for preventive chronic disease in human being[J], Evid.-Based Com. Alt. Med., 2017, 2017, ID94028849, 1-13.

[88] Kuok Q.Y., Yeh C.Y., Su B.C., Hsu P.L., Ni H., Liu M.Y., The triterpenoids of Ganoderma tsugae prevent stress-induced myocardial injury in mice[J], Mol. Nutr. Food Res., 2013,57(10), 1892-1896.

[89] Zhang S., He B., Ge J., Zhai, C., Liu X., Liu P., Characterization of chemical composition of *Agaricus brasiliensis* polysaccharides and its effect on myocardial SOD activity, MDA and caspase-3 level in ischemia-reperfusion rats[J], Int. J. Biolog. Macromol., 2010,46(3), 363-366.

[90] Monteiro C.S., Miguel O.G., Eugênia B.M., Penteado P.T., Haracemiv S.M., Extraction of lycopene from tomato sauce with mushrooms (*Agaricus brasiliensis*), determined by high - performance liquid chromatography[J], Int. J. Food Sci. Nutr., 2009, 1,72-78.

[91] Bisen P.S., Baghel R.K., Sanodiya B.S., Thakur G.S., Prasad. G.B., 2010.*Lentinus edodes*: a macrofungus with pharmacological activities[J], Curr. Med. Chem., 2010, 17(22), 2419-2430.

[92] Zhou X., Gong Z., Su Y.; Lin J., Tang K., Cordyceps fungi: natural products, pharmacological functions and developmental products[J],. J. Pharm. Pharmacol., 2009,61(3), 279-291.

[93] Nguyen H.N., Fujiyoshi A., Abbott R.D., Miura K., Epidemiology of cardiovascular risk factors in Asian countries[J], Circul. J., 2013,77(12), 2851-2859.

[94] Ueshima H., Explanation for the Japanese paradox: prevention of increase in coronary heart disease and reduction in stroke[J], J. Atherosclerosis Thrombosis, 2007.14: 278–286.

[95] Hu E.A., Pan A., Malik V., Sun Q., White rice consumption and risk of type 2 diabetes: meta-analysis and systematic review[J], Br. Med. J.,2012,344,e1454.

[96] Zeng Y.W., Sun D., Du J., Pu X.Y., Yang S.M., Yang X.M., et al., Identification of QTLs for resistant starch and total alkaloid content in brown and polished rice[J], Genet. Mol. Res., 2016, 15(3), 10.4238/gmr. 15037268.

[97] Zhou H., Wang L., Liu G.,Mang X., Jing Y., Shu X., et al., Critical roles of soluble starch synthase SSIIIa and granule-bound starch synthase Waxy in synthesizing resistant starch in rice [J], Proc. Nat. Acad. USA, 2016,113(45),12844-12849.

[98] Dai F., Nevo E., Wu D.Z., Comadran J., Zhou M.X., Qiu L.,et al., Tibet is one of the centers of domestication of cultivated barley[J], Proc. Nat. Acad.Sci. USA, 2012,109 (42), 16969 -16973.

[99] Talhelm T., Zhang X., Oishi S., Shimin C., Duan D., Lan X., Large-scale psychological differences within China explained by rice versus wheat agriculture[J], Science, 2014,344(6184), 603-608.

[100] Wynn J.G., Sponheimer M., Kimbel W.H., Alemseged Z., Reed K., Bedaso Z.K., et al., Diet of *Australopithecus afarensis* from the Pliocene Hadar Formation, Ethiopia[J], Proc. Nat. Acad. Sci. USA,2013,110 (26), 10495-10501.

[101] Venugopal S., Iyer U.M., Management of diabetic dyslipidemia with subatmospheric dehydrated barley grass powder[J], Int. J.Green Pharm., 2010,4, 251-256.

[102] Cerling T.E., Chritz K.L., Jablonski N.G., Leakey M.G., Manthi F.K., Diet of *Theropithecus* from 4 to 1 Ma in Kenya[J], Proc. Nat. Acad. Sci. USA,2013, 110(26), 10507-10512.

F.X. Wang, F. Pan, S. Zhang, C.M. Liu, R.Y. Li, J.W. Liang, Y. Wan, G.M. Fu*

Determination of Glyphosate Residue in Genetically Modified Soybean by Protein Precipitator Clean-up and HPLC with OPA Post-column Derivatization

Abstract: The method of detecting glyphosate in genetically modified (GM) soybean through High Performance Liquid Chromatography (HPLC) with post-column derivatization was established and chromatography conditions were optimized in this paper. Three precipitators, such as acetonitrile, hydrochloric acid and a combination of potassium ferrocyanide and zinc acetate, were used respectively to remove protein from soybean slurry. Results show that the purifying effect of hydrochloric acid was better than others. Then the precipitated soybean solution sample after centrifugation was defatted with methylene chloride, and the supernatant was analyzed by HPLC with post-column derivatization. Sample recovery rates are between 80% and 93%. The relative standard deviations (R.S.D.) of the glyphosate content measurements are in the range 2.87-4.98%, and the detection limit is 2.5 μg/mL. So this method has high sensitivity and accuracy, and it could be applied in detecting glyphosate residue in other protein grains.

Keywords: genetically modified soybean, clean-up, glyphosate, HPLC.

1 Introduction

In developed countries, the great economic benefit had been taken with the planting of GM crops while the GM soybean was the most important one of glyphosate-resistant crops. Glyphosate, called N-(methyl phosphonic) glycine, was regarded as a kind of low toxicity, high-efficiency, broad-spectrum herbicide introduced by the Monsanto Company in the United States. It was reported that glyphosate could be transferred and gathered in fishes, amphibious animals, causing vertebrates morphological changes of organs [1,2]. Research of WHO showed that keeping in touch with glyphosate for a long time could induce various symptoms, such as the decrease of mice fetal number

***Corresponding author: G.M. Fu,** State Key Laboratory of Food Science and Technology, Nanchang University, Nanchang, China, E-mail: fuguiming@ncu.edu.cn
F.X. Wang, F. Pan, C.M. Liu, R.Y. Li, Y. Wan, State Key Laboratory of Food Science and Technology, Nanchang University, Nanchang, China
S. Zhang, Jiangxi Provincial Product Quality Supervision Testing College, Nanchang, China
J.W. Liang, School of food science and technology, Nanchang University, Nanchang, China

and sperm, the increasing rate of preterm birth and the promoting of abortion [3]. And Thongprakaisang et al found in 2013 that the presence of even very low concentration of glyphosate (10^{-12}-10^{-6} mol/L) could facilitate the growth of cancer cells [4]. In recent years, the residue of glyphosate and its metabolite aminomethyl phosphoric acid (AMPA) had been detected in GM soybean [5], where the content of glyphosate residue was 3.908 mg/kg and AMPA was 3.364 mg/kg. The limited content of glyphosate in soybean prescribed by US, European Union and Japan was 20 mg/kg [6]. As the largest GM soybean importer, China's soybean imports reached 63.38 million tons in 2013. More than 90% of soybean oil and blend oil containing soybean ingredient were made from GM soybean [7]. In China, with the increasing consumption and import of GM soybean, edible risk brought by the glyphosate residue in soybean was gradually enhanced. Therefore, surveillance of the glyphosate residue in GM soybean should be strengthened. Beyond that, establishing a high-efficiency and sensitive method to determinate glyphosate was an extremely urgent problem.

Hardly was glyphosate dissolved in common organic solvents and its solubility was only 1.2% in water at room temperature. Glyphosate, having chemical and physical properties of strong polarity, absence of chromophore and fluorophore, was difficult to volatilize [8]. In addition, the high content of oil and protein in soybean would interfere with UV absorption of glyphosate. Therefore, it was a challenge to detect the glyphosate accurately. At present, several methods of detecting the residual glyphosate in drinkable water, fruits and vegetables have been described. For example, glyphosate of Oolong tea, extracted through ultrasonic and solid phase extraction (SPE) technique, was detected by HPLC-MS after derivatization [9], but the equipment was expensive, and that is why its popularity was relatively low. Glyphosate residue of apple, which was extracted with deionized water and purified by ion exchange columns, was detected by GC-NPD after derivatization [10], but steps were tedious and the cost was higher. Those methods above suffered from complicated and tedious sample preparation. Glyphosate content of poisoned blood was observed by UV spectrophotometry [11]. However, the UV spectrophotometry method has poor reproducibility, more disturbed factors and low accuracy.

To obtain a suitable detection method for detecting glyphosate residue in GM soybean rich in protein and fat, HPLC method was developed in this paper. Methods of HPLC with pre-column derivatization, UV detection and post-column derivatization fluorescence detection were investigated respectively, expecting to establish an accurate and reliable method fot he detection of glyphosate residue in GM soybean.

2 Experimental

2.1 Reagents

Glyphosate standard substance was purchased from Dr. Ehrenstorfer GmbH (Germany). Transgenic soybean inputted from Columbia was provided by Nantong Laibao Oil Plant. Lithium Eluant (Cat.NO:1700-1125), O-Phthalic Aldehyde (OPA) (Cat. NO:OD104), OPA diluent (Cat.NO:O120), and THIOFLUOR™ (Cat.NO:3700-2000) were obtained from Pickering Laboratories Company (California, United States). Methanol, methylene dichloride, and acetonitrile were chromatographic grade reagents from Honeywell B&J (Morris County, New Jersey, United States). Hydrochloric acid, ferrocyanide and zinc acetate were domestic analytical reagents from Xilong Chemical Co., Ltd. (Guangdong, China). Other reagents were analytical pure. Deionized water obtained from a MilliQ water purification system (Millipore Ltd., Massachusetts, United States) was used for solution preparation.

2.2 Instrumentation

An Agilent LC-1100 HPLC (Agilent Technologies Inc., California, United States) with an online degasser, a quaternary pump, fluorescence detection (FLD) system, an auto sampler, and post-column derivatization system (Pickering Laboratories Company, California, United States).

2.3 Sample Preparation

The powdered GM soybean (5 g) was weighed thrice as three samples, placed in centrifuge tube with 30 mL deionized water and dealt with ultrasound for 30 min. Then three cleaning agents, such as 5 mL acetonitrile [12], 5 mL potassium ferrocyanide (106 g/L), and the combination of 5 mL zinc acetate (220 g/L) and 200 μL hydrochloric acid, were added into the sample solution respectively [13]. After the sample solution was diluted to 50 mL with deionized water and processed by ultrasound for 10 min, the supernatant was gotten through centrifugation (10,000 rpm) for 5 min at 0°C. Next, 10 mL supernatant was mixed with 15 mL methylene dichloride [13], vortex-mixed for 10 min and centrifuged (10,000 rpm) for 5 min at 0°C. The upper centrifugal liquid was placed to a new container and the described step was repeated twice. Finally, the purified extract liquid for HPLC determination, was prepared.

2.4 Detection of Glyphosate Content by HPLC with Pre-column Derivatization

The sample was tested using the method of nitrosation in this paper. The purified extract liquid from Step 2.3 (5 mL) was added with 2 mL sulfuric acid (100 g/L) and 0.5 mL potassium bromide (2.0 mol/L), shaken up and added with 0.5 mL sodium nitrite (50 g/L), followed by resting place for 30 min. The volume was adjusted to 100 mL with distilled water, and the sample was analyzed through HPLC after being filtrated through 0.45 μm membrane.

The analysis column was ZORBAX Eclipse XDB-C_{18} (150 mm×4.6 mm, 5 μm, Agilent Technologies Inc., California, United States). Methanol (A) and water (B) (v (methanol): v (water) = 2:98, 0.1% phosphoric acid was contained in water phase) was used as the mobile phase at the constant flow rate of 0.8 mL/min. The UV detective wavelength was 240 nm. And the column temperature was maintained at 25°C.

2.5 Detection of Glyphosate Content by HPLC with Post-column Derivatization

The sample determination was performed using an HPLC-FLD system. OPA solution used for post-column derivatization was prepared before use as follows. OPA (O-Phthalaldehyde) (63 mg) was dissolved with 2 mL methanol followed by addition of 0.42 g THIOFLUOR™. The liquid volume was adjusted to 200 mL with OPA diluent and filtrated through 0.45 μm membrane.

The analysis column was Thermo LITHIUM AMINO ACID ANALYSIS chromatography column (100 mm×4.0 mm, 5 μm, Picking Laboratories United States) and the column temperature was maintained at 40°C. Lithium Eluant was used as the mobile phase at flow rate of 0.3 mL/min. The FLD detector with wavelengths was set at 355 nm (Ex) and 465 nm (Em) and the reaction tank temperature was maintained at 44°C.

2.6 Optimization of Chromatographic Conditions and Validation of the Method

Good purification and derivatization methods were chosen in order to detect and analyze glyphosate. And the chromatographic conditions and method effectiveness were observed.

1. **Optimization of chromatographic conditions:** The glyphosate standard liquid (400 μg/mL) chromatography spectra taken respectively by the different elution flow rate of 0.2 mL/min [14] and 0.3 mL/min [9], and the different wavelength of 355 nm (Ex) [15] and 330 nm (Ex) [16] were compared to choose the suitable flow rate and detective wavelength.
2. **Validation of the method:** Four batches of spiked samples were analyzed to evaluate the performance of the method. The smashed soybean (5 g) was weighed

accurately and placed into 50 mL polypropylene centrifuge tube. The spiked levels were 5 μg/mL, 10 μg/mL, 15 μg/mL, and 20 μg/mL (n=4). The method was valid when the R.S.D was less than 5%.

3 Results and Discussion

3.1 The Purification Method

Samples disposed with different precipitators were detected by HPLC with post-column derivatization at the flow rate 0.3 mL/min and the wavelength 355 (Em) and 465 (Ex) (Fig. 1-Fig. 3). Results showed that the purifying effect of hydrochloric acid was the best because the retention time of glyphosate standard (400 μg/mL) was 3.8 min and the impurity peaks between 3.5 min and 5.0 min were relatively weak, the better was acetonitrile precipitation. Considered hydrochloric acid was the cheapest among three precipitators, this reagent was chosen to purify soybean slurry, mainly for removing protein from the sample. Because some interference could be still existed in the disposed sample (Fig. 3), dichloromethane was adopted to purify the sample further. A large amount of lipid was removed by dichloromethane and the interference to instrument was avoided, though the HPLC spectrum was not changed notably (Fig. 4). This step greatly reduced the risk of damaging the equipment.

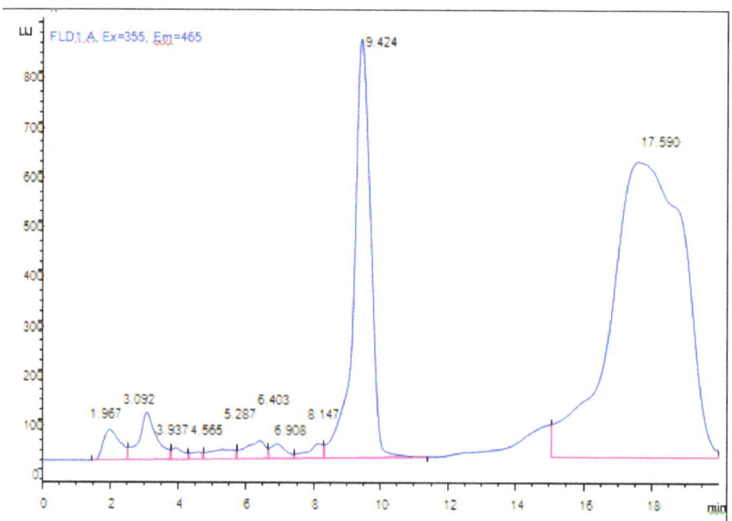

Figure 1: Chromatogram obtained from sample with acetonitrile as protein precipitator.

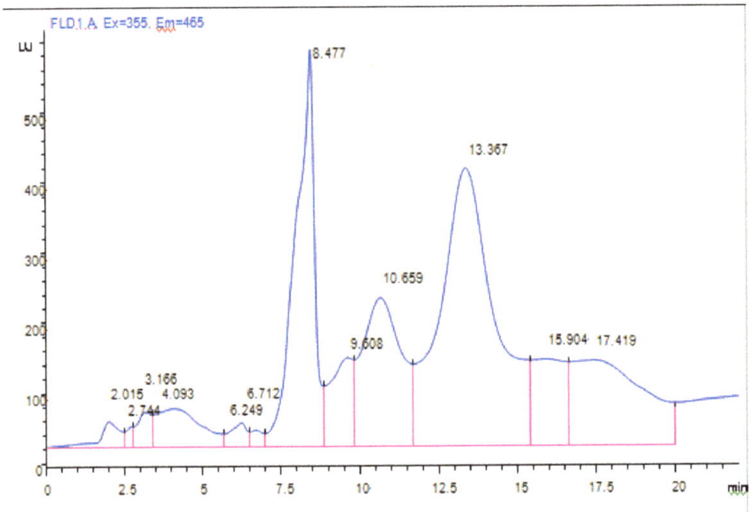

Figure 2: Chromatogram obtained from sample with ferrocyanide and zinc acetate as protein precipitator

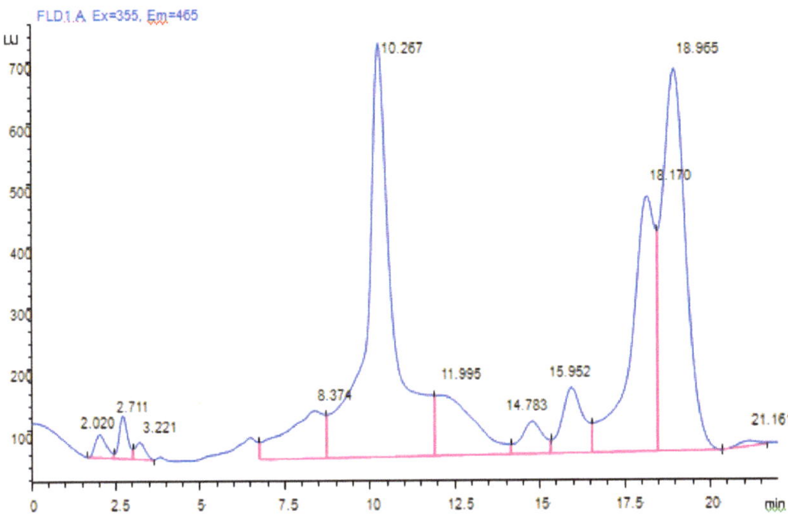

Figure 3: Chromatogram obtained from sample with hydrochloric acid as protein precipitator.

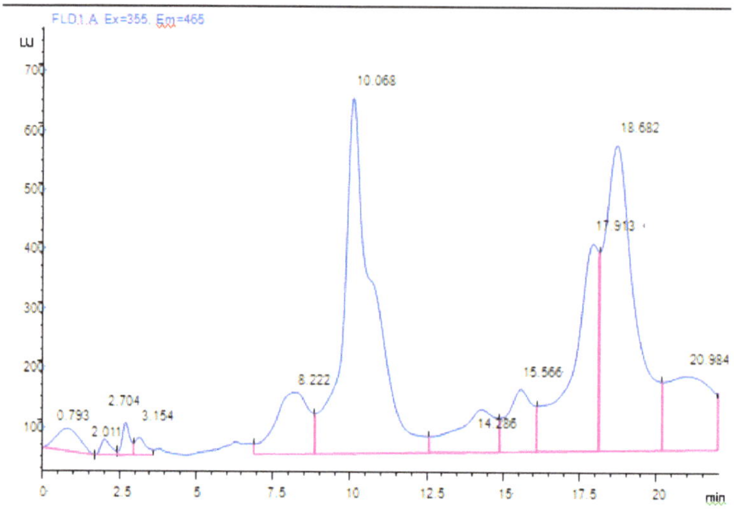

Figure 4: Chromatogram obtained from sample purified by dichloroethane with hydrochloric acid as protein precipitator.

3.2 The Derivatization Method

The spectra of glyphosate standard (400 μg/mL) by HPLC with pre-column derivatization and by HPLC with post-column OPA derivatization were illustrated in Fig. 5 and Fig. 6. The HPLC with pre-column derivatization method has been applied in determinating biogenic amines in germinated and fermented brown rice [17], and glyphosate in water and plant material [18]. Characteristic peak of glyphosate that appeared at 5.5 min was weak (Fig. 5). An interference peak at 4.8 min which was possibly the product from nitrosation derivatization had almost the same response value as glyphosate and must be a great influence on detecting glyphosate. Meanwhile, it showed that the impurity peak at 2.1 min in the spectrum from HPLC with post-column OPA derivatization (Fig. 6) was weak so its effect on the strong characteristic peak of glyphosate at 3.8 min could be neglected. Therefore, the latter method was more suitable to detect glyphosate abstracted from GM soybean than the former. It was reported that N-methylcarbamate pesticides in vegetable and water samples [19], and organothiophosphorus pesticides in water could be determined by HPLC with post-column chemiluminescence detection [20].

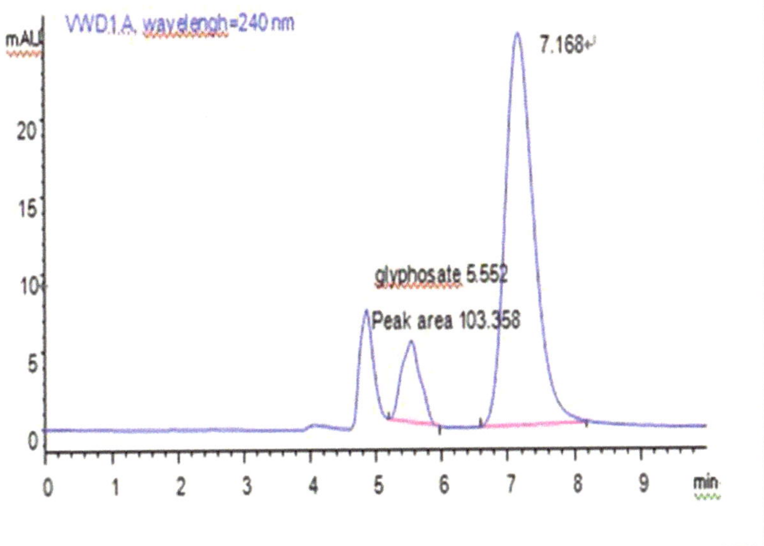

Figure 5: The chromatogram of glyphosate by HPLC with pre-column derivatization.

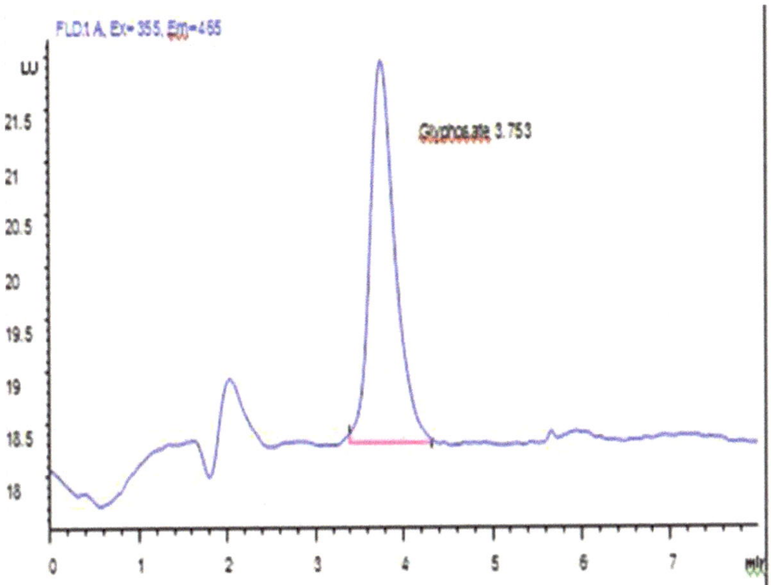

Figure 6: The chromatogram of glyphosate standard by HPLC with OPA post-column derivatization.

3.3 Optimization of HPLC post-column OPA Derivatization Chromatographic

1. **The flow rate:** The glyphosate standard was detected at the flow rate of 0.2 mL/min and 0.3 mL/min respectively (Fig. 7 and Fig. 8). The retention time of glyphosate peak was 5.45 min and peak shape was wide, the impurity peaks appeared before glyphosate had serious trailing phenomenon at the flow rate of 0.2 mL/min (Fig. 7). In contrast, the peak shape of glyphosate at the flow rate of 0.3 mL/min was sharp and kept a certain distance with impurity peaks, which reduced the interference (Fig. 8). So the flow rate 0.3 mL/min was chosen.

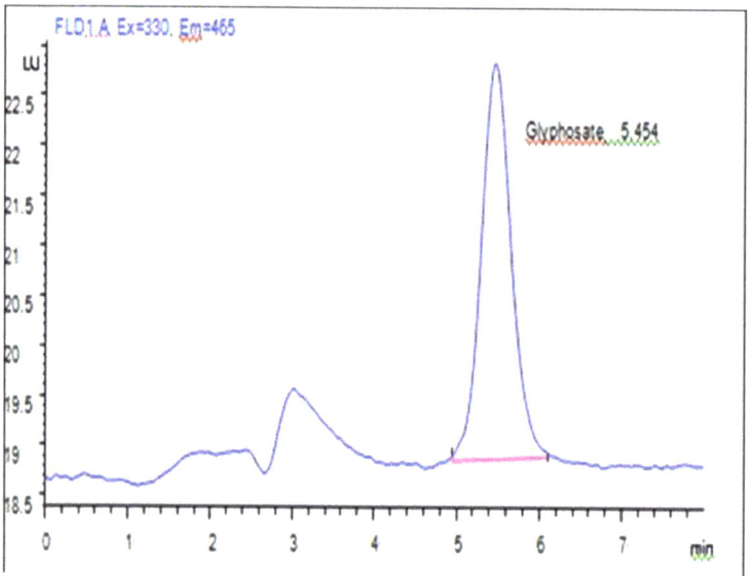

Figure 7: The chromatogram of glyphosate standard at the flow rate of 0.2 mL/min (the glyphosate concentration of 400 μg/mL)

2. **The wavelength:** The glyphosate standard (400 μg/mL) was detected at the wavelength of 330 nm (Ex) and 355 nm (Ex) respectively, which's peak area was 71.9 mAU*s and 86.7 mAU*s (Fig. 8 and Fig. 6). The response value of glyphosate at 355 nm (Ex) was higher, so the 355 nm (Ex) wavelength was chosen.

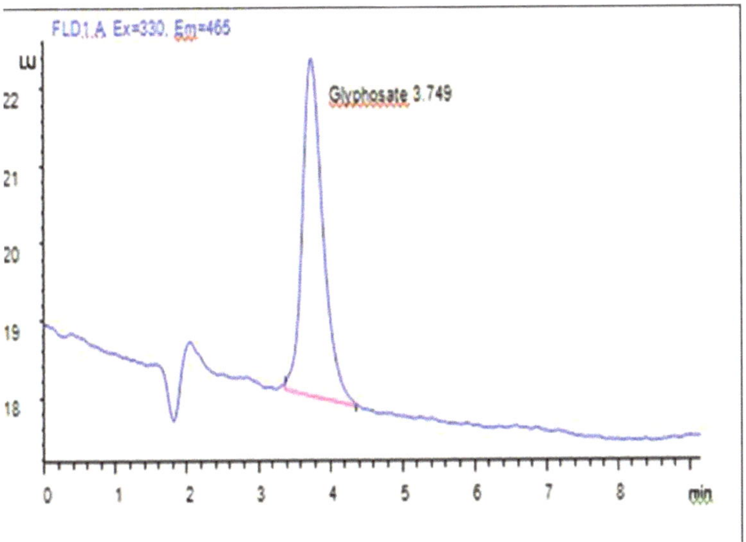

Figure 8: The chromatogram of glyphosate standard by HPLC with OPA post-column derivatization.

3.4 Detection Limit

A series of glyphosate standard samples were analyzed according to the optimized condition of HPLC. The regression statistics were calculated from the calibration curve constructed. The equation of linear regression was y=0.181x-0.440 and showed satisfactory linearity with the correlation coefficient (r2=0.9999) greater than 0.999. There was a good linear relationship in the range of 0.5-4.0 μg/mL.

The sensitivity of the method was expressed as limit of detection (LOD). The LOD was 2.5 μg/mL, calculated as three times the average baseline noise (S/N=3).

3.5 Recovery and Precision

The recovery and precision of the method were evaluated with adopting optimized purification condition and listed in Table 1. Sample recovery rates were 80.7-92.8%, R.S.D. of the glyphosate content measurement were 2.87-4.98%, which was less than 5%. So this method was high repeatability.

Overall, although the HPLC method with pre-column derivatization has been successfully applied for detecting glyphosate residue in chestnut which was obtained by extracting with water and purifying with SPE [21], the application in determining glyphosate residue in soybean was not as easy as expected.

Table 1: Recovery rates and precisions of the method

Sample No.	Matrix value (μg/mL)	Spiked concentration (μg/mL)	Recovery rate (%)	Relative standard Deviations (%)
1	n.a	5	80.7	4.98
2	n.a	10	82.3	4.48
3	n.a	15	85.9	3.49
4	n.a	20	92.8	2.87

Results showed that the glyphosate peak from pre-column derivatization was disturbed by impurity, while the glyphosate peak from post-column derivatization showed a good shape and was hardly interfered by impurity. Up to now, there have been several methods used in detecting residual glyphosate in food. HPLC-MS method was applied for detecting glyphosate residue in rice, maize and soybean samples which was ultrasonic extracted and purified with SPE [22]. Soybean sample for HPLC-MS analysis was prepared after methanol deproteinization and methylene chloride decontamination [23]. The glyphosate in fruits and vegetables was detected by mixed-mode hydrophilic interaction/weak anion-exchange liquid chromatography couples with electrospray tandem mass spectrometry [24]. Compared with above mentioned methods, the method of HPLC with post-column OPA derivatization could directly detect glyphosate residue in samples, low cost, high efficiency.

3.6 Sensitivity and High Popularity.

In this experiment, there was no residual glyphosate detected in Columbia soybean sample by HPLC with post-column derivatization under optimized conditions. This might be a reason that the glyphosate residue was less in the soybean sample used in the experiment and lower than LOD of equipment. So this batch of transgenic soybean inputted from Columbia was safe in its glyphosate content. Mo et al tested residual glyphosate in sugarcane sample extracted with acetonitrile by UPLC-MS [25]. The result suggested that residual glyphosate of samples was not being detected. Li et al. used hydrochloric acid and methylene chloride to deal with protein in soybean samples, and analyzed it by GC-MS after purifying by anion-exchange column [14]. The result was that glyphosate residue of non-GM soybean was not detected and glyphosate residue content in GM soybean was 0.08 mg/kg.

4 Conclusions

In this paper, the method of detecting glyphosate in soybean was developed. Three different precipitators, acetonitrile, potassium ferrocyanide, and the combination of zinc acetate and hydrochloric acid, were compared to choose a suitable purifying soybean slurry reagent to remove protein. Among them, hydrochloric acid had the best purification efficiency. The HPLC with OPA post-column derivatization method had more advantage than HPLC with pre-column derivatization method. The optimized conditions were the flow rate 0.3 mL/min and the wavelength 355 nm (Ex) and 465 nm (Em). This method was developed and validated for detecting glyphosate residue in soybean. In the concentration range of 0.5-4.0 μg/mL, the correlation coefficient (r^2) of glyphosate standard curve was 0.9999, the sample recovery rates were 80.7-92.8%, R.S.D. of the glyphosate content measurements were 2.87-4.98%, and the detection limit was 2.5 μg/mL. This method, as an attractive alternative to the traditional or expensive methods, allowed a rapid and direct determination of glyphosate residue in GM soybean and is expected to be applied in other high protein grains.

Acknowledgment: This work was funded by the National Natural Science Foundation of China under Grant No. 31160336. The authors were grateful to all the students for their valuable suggestions as well as technical support.

Conflict of Interest: This article does not contain any studies with animal or human subject.

References

[1] Junges C. M., Vidal E. E., Attademo A. M., Mariani, M. L., Cardell, L., Negro, A. C., et al. Effectiveness evaluation of glyphosate oxidation employing the H_2O_2/UVC process: Toxicity assays with Vibrio fischeri and Rhinella arenarum tadpoles[J]. J. Environ. Sci. Health. Part B., 2013, 48(3): 163-170.

[2] Relyea R. A. New effects of Roundup on amphibians: Predators reduce herbicide mortality; herbicides induce antipredator morphology[J]. Ecol. Appl., 2012, 22(2): 634-647.

[3] Mañas F., Peralta L., Raviolo J., Ovando, H. G., Weyers, A., Ugnia, L., et al. Genotoxicity of AMPA, the environmental metabolite of glyphosate, assessed by the Comet assay and cytogenetic tests[J]. Ecotoxicol. Environ. Safety, 2009, 72(3): 834-837.

[4] Thongprakaisang, S., Thiantanawat, A., Rangkadilok, N., Suriyo, T., Satayavivad, J., et al. Glyphosate induces human breast cancer cells growth via estrogen receptors[J]. Food Chem. Toxicol., 2013, 59: 129-136.

[5] China Food Economic Net. Determination of pesticide residue in imported genetically modified soybean products[J]. Chinese Fruits Vegetables. 2014(7): 16.

[6] Yang M.S., Wang R.Q., Yuan Y.W., Cui Y.H., Zhou Y.L., Consider standard system and quality safe in China from glyphosate residue in soybean[J]. Qual. Safe Agr. Prod, 2011 (3): 14-16.

[7] Yang G.S., Research of China's soybean industry economic under industry chain[D]. China Agr. Univer. Econ. Manage Acadeny, 2014.

[8] Schrübbers, L. C., Masís-Mora, M., Rojas, E. C., Valverde, B. E., Christensen, J. H., Cedergreen, N. Analysis of glyphosate and aminomethylphosphonic acid in leaves from Coffea arabica using high performance liquid chromatography with quadrupole mass spectrometry detection[J]. Talanta, 2016, 146: 609-620.
[9] Chen L., Determination of glufosinate-ammonium, glyphosate and ammonia methyl phosphonic acid residues in Oolong tea by HPLC-MS/MS[J]. Tea Sci. Tech. 2014(1): 21-27.
[10] Hu J.Y., Zhao D.Y., Ning J., Chen C.L., Li J. Z., Detection of glyphosate residue in soil and apple by GC-nitrogen phosphorus (NP) detecto[J]. J Pestici Sci. 2007, 9(3): 285-290.
[11] ZhaoS.Y., Li S.Q., Du L.X., Li F.L., Determination of glyphosate poisoning blood with ultrsviolet spectrophotometry method[J]. Labeled Immun Clinic, 2014, 21(3): 351-354.
[12] Wei, X., Gao, X., Zhao, L., Peng, X., Zhou, L., Wang, J., Pu, Q. Fast and interference-free determination of glyphosate and glufosinate residues through electrophoresis in disposable microfluidic chips[J]. J. Chromatogr. A, 2013, 1281: 148-154.
[13] Li H. M., Wang L. J., Yang Y., Wang X., Xu R.Y., Yang M. M., et al. The analysis experiment report of glyphosate in drinking water by HPLC with post-column derivatization. The northwest region the seventh chromatographic academic report, GanSu province the 12th conference proceedings. 2012.
[14] Li B., Guo D. H., Zhu J., Han L., Yang J. X., Determination of glyphosate and aminomethylphosphonic acid residues in plant by GC-MS[J]. J Instru. Anal, 2005, 24: 345-348.
[15] Chen J., Yuan X. J., Hu J. Q., Zhan Z. K., Detection of glyphosate in water by HPLC with post-column derivatization[J]. Chemical Enterprise Management. 2014: 43-45.
[16] Zhi J. L., Mu R. X., Chen M. X., Zhu Z. W., Determination of glyphosate and aminomethylphosphonic acid residua in food by HPLC with post-column[J]. China Anal Lab, 2008, 27: 170-172.
[17] Hayat, A., Jahangir, T. M., Khuhawar, M. Y., Alamgir, M., Hussain, Z., Haq, F. U., Musharraf, S. G. HPLC determination of gamma amino butyric acid (GABA) and some biogenic amines (BAs) in controlled, germinated, and fermented brown rice by pre-column derivatization[J]. J. Cereal Sci., 2015, 64: 56-62.
[18] Nedelkoska, T. V., Low, G. C. High-performance liquid chromatographic determination of glyphosate in water and plant material after pre-column derivatisation with 9-fluorenylmethyl chloroformate[J]. Anal. Chim. Acta, 2004, 511(1): 145-153.
[19] Huertas-Pérez, J. F., García-Campaña, A. M. Determination of N-methylcarbamate pesticides in water and vegetable samples by HPLC with post-column chemiluminescence detection using the luminol reaction[J]. Anal. Chim. Acta, 2008, 630(2): 194-204.
[20] Catalá-Icardo, M., Lahuerta-Zamora, L., Torres-Cartas, S., & Meseguer-Lloret, S. Determination of organothiophosphorus pesticides in water by liquid chromatography and post-column chemiluminescence with cerium (IV)[J]. J. Chromatogr. A, 2014, 1341: 31-40.
[21] Zhou Y. M., Li N., Zhao Y. B., Zang J., Research of HPLC method for detecting glyphosate residue in chestnut. Food Sci. 008, 29(9): 461-464.
[22] Botero-Coy, A. M., Ibáñez, M., Sancho, J. V., Hernández, F Direct liquid chromatography–tandem mass spectrometry determination of underivatized glyphosate in rice, maize and soybean[J]. J. Chromatogr. A, 2013, 1313: 157-165.
[23] Martins-Júnior, H. A., Lebre, D. T., Wang, A. Y., Pires, M. A., Bustillos, O. V. An alternative and fast method for determination of glyphosate and aminomethylphosphonic acid (AMPA) residues in soybean using liquid chromatography coupled with tandem mass spectrometry[J]. Rapid Commun. Mass Sp., 2009, 23(7): 1029-1034.
[24] Chen, M. X., Cao, Z. Y., Jiang, Y., & Zhu, Z. W. Direct determination of glyphosate and its major metabolite, aminomethylphosphonic acid, in fruits and vegetables by mixed-mode hydrophilic interaction/weak anion-exchange liquid chromatography coupled with electrospray tandem mass spectrometry[J]. J. Chromatogr. A, 2013, 1272: 90-99.
[25] Mo J. L., Miao L., Gan N. J., Determination of glyphosate in sugar cane by UPLC-MS/MS[J]. Mod Food Sci. Tech. 2011, 27(9): 1143-1146.

Mingyue Li, Hong Yang*

Preliminary Study on Biological Effects of MiR-702-3p in Epithelial–mesenchymal Transition Induced by Nano-SiO_2 Particles

Abstract: Nano-sized SiO_2 particles are gradually applied in numerous fields, the biosecurity of nano-sized SiO_2 particles remains to be controversial. The objective of this study was investigating the biological effects of miR-702-3p in epithelial–mesenchymal transition (EMT) induced by nano-SiO_2 particles. In our previous study, results from Illumina Hiseq2000 high throughput sequencing has shown that miR-702-3p is an up-regulated miRNA in fibrotic pulmonary tissue induced by nano-SiO_2 particles on day 60 and 90. In this study, rats' lung tissue treated by nano-SiO_2 particles of 12.5 mg/ml on day 60 was chosen for the research while the control group was rats' lung tissue treated by equivalent saline. Target genes of miR-702-3p were predicted using the Target Scan (http://www.targetscan.org/). With the analysis of Gene Ontology and KEGG, BMP-7 was filtrated to be the target genes of miR-702-3p for further study. Dual-Luciferase Reporter Assay was used to verify the relationship betweenmiR-702-3p and BMP-7. The expression level of miR-702-3p and BMP-7 in pulmonary fibrosis induced by nano-SiO_2 particles was detected by Quantitative real-time PCR and western blot analysis respectively. To analyze the correlation of EMT and pulmonary fibrosis induced by nano-SiO_2 particles, the expressions of TGF-β1, Smad3, E-cad, α-SMA and COL3 were determined by Quantitative real-time PCR and western blot analysis respectively. Result of Dual-Luciferase Reporter Assay indicated that miR-702-3p directly regulates the post-transcriptional manner of BMP-7.The expression of miR-702-3p in the lung tissue of nano-SiO_2 particles group was up-regulated (fold changes > 2.0) when compared with the saline group, which was consistent with the result of Illumina Hiseq2000 high throughput sequencing. The expression of BMP-7 mRNA and protein in the lung tissue of nano-SiO_2 particles group was down-regulated (fold changes < 0.5) when compared with the saline group ($P<0.05$). The expression of TGF-β1, Smad3, α-SMA, and COL3 were up-regulated ($P<0.05$) compared with the saline group both at mRNA and protein level. E-cad mRNA and protein were both down-regulated ($P<0.05$). In conclusion, miR-702-3p may be a positive regulator in pulmonary fibrosis induced by nano-SiO_2 particles, and BMP-7 was a target gene of miR-702-3p in regulating pulmonary fibrosis. MiR-

***Corresponding author: Hong Yang,** Key Laboratory of Environmental Medicine Engineering, Ministry of Education, School of Public Health, Southeast University, Nanjing 210009, Jiangsu, China, E-mail: yanghongr@seu.edu.cn

Mingyue Li, Key Laboratory of Environmental Medicine Engineering, Ministry of Education, School of Public Health, Southeast University, Nanjing 210009, Jiangsu, China

702-3p may directly weaken the post-transcriptional manner of BMP-7 to promote the process of EMT which contributes to pulmonary fibrosis induced by nano-SiO_2 particles.

Keywords: component, MiR-702-3p, pulmonary fibrosis, epithelial-mesenchymal transition, Nanosized SiO_2, Bone morphogenetic protein-7.

1 Introduction

Silicosis is characterized by the silicosis nodules formation and development of pulmonary interstitial fibrosis with progressive impairment of pulmonary function which is caused by the persistent inhalation of silica particles [1,2]. Nano-sized SiO_2 particles are gradually applied in numerous fields like food industry, coating, rubber, cosmetics and pharmaceutical and health care and so on due to its special physical and chemical properties. Nano-SiO_2 particles have the same chemical composition as silica particles. Silica particles can lead to lung diseases such as silicosis and lung cancer [3,4], but the correlation of nano-SiO_2particles and pulmonary fibrosis is still controversial. In recent years studies have shown that nano-SiO_2particles can cause pulmonary fibrosis after sub-chronic or chronic exposure. Tracheal instillation of nano-SiO_2 particles in rats can lead to inflammation for 7 days and cause pulmonary fibrosis for 30 days [5]. Mice after pulmonary perfusion of nano-SiO_2 particles demonstrated significant lung inflammation and end up with the occurrence of pulmonary fibrosis [6]. Rats sacrificed at day 28 after exposure under continuous dynamic inhalation conditions presented with pulmonary fibrosis formation [7]. Our previous studies have also verified that rats treated with nano-SiO_2particles through endotracheal instillation resulted in pulmonary injury, which mainly manifested in pulmonary inflammation at protophase (7d, 15d, 30d) and pulmonary interstitial fibrosis at anaphase (60d and 90d) [8].

The epithelial–mesenchymal transition (EMT) is a prevailing phenomenon that plays a crucial role in organic fibrosis and cancer metastasis [9]. EMT is a process of trans-differentiation from the epithelial cells of mature differentiation into the mesenchymal cells of complete differentiation, mainly transforming into fibroblasts and myofibroblasts [10]. During this process, epithelial cells will lose their markers, such as E-cadherin (E-cad) and cytokeratin, as well as mesenchymal cells acquisition of markers, such as vimentin and a-smooth muscle actin (a-SMA) [11,12], meanwhile a large number of extracellular matrix (including COL1 and COL3) was secreted. EMT is a critical process in pulmonary fibrosis. TGF-β1 is a member of the transforming growth factor superfamily, which can be secreted by neutrophils, eosinophils, macrophages, fibroblasts and myofibroblasts, and studies have reported that TGF-β1 plays an important role in regulating the process of EMT [13,14]. TGF-β signaling has been reported to be mediators of EMT, induced by TGF-β through Smad2/3 [15,16]. TGF-β1 can promote the formation of fibrosis by stimulating fibroblast translate into myofibroblasts and secreting a large amount of collagen to make the extracellular matrix depositing [17] and inducing the EMT process [16]. The canonical Smads-dependent pathway is one of the main pathways in the process of TGF-β1-induced EMT. [18]

With the development of epigenetics, microRNAs (miRNAs) are gradually attracting more and more attention on account of their extensive regulating effect in organism. MiRNAs are short non-coding endogenous RNAs of approximately 23 nucleotides that mainly modulate gene expression through facilitating degradation of target mRNAs or suppressing translation by binding to the 3'Untranslated Region (UTR) [19,20]. MiRNAs have the characteristics of specificity of time and change of miRNAs occurs before the pathological change of a disease, based on gene expression profiling technology it provides new ideas and methods for the research of pulmonary toxicity effect and mechanism of nano-SiO_2 particles. MiRNAs play a momentous regulatory role in the process of cell apoptosis, tissue inflammation and fibrosis [21-23]. In recent years, some studies have reported that miRNAs play crucial roles in regulating the development of pulmonary fibrosis [24-28]. Results from Illumina Hiseq2000 high throughput sequencing in our previous study have shown that miR-702-3p is a dysregulated miRNA that up-regulated in fibrotic pulmonary tissue at 60 and 90 days [29]. Present studies have reported that miR-702 have the function of promoting the proliferation of embryonic stem cell [30] and inhibiting cell apoptosis [31]. Results of gene ontology analysis showed that miR-702-3p might participate in the development process of pulmonary fibrosis through regulating TGF-β signaling pathway, formation of connective tissue, smad-dependent pathway, metabolism of collagen, wnt receptor signaling pathway. It is speculated that miR-702-3p may play an important role in the development process of pulmonary fibrosis induced by nano-SiO_2 particles. Prediction of the database targetscan.org displayed that, bone morphogenetic protein 7 (BMP-7) might be the downstream target genes of miR-702-3p.BMP-7 belonging to the TGF-β superfamily which is a negative regulator of pulmonary fibrosis. BMP-7 can reverse the process of pulmonary fibrosis by antagonizing TGF-β/Smad signaling pathway [32] and inhibiting EMT [33]. Considering the above, we assume that miR-702-3p and its target genes BMP-7 may contribute to pulmonary fibrosis induced by nano-SiO_2 particles through regulating EMT. In this study, we investigate the biological effects of miR-702-3p and the correlation of miR-702-3p and EMT in pulmonary fibrosis induced by nano-SiO_2 particles.

2 Material and Methods

2.1 Nano-sized SiO_2 Particles and Animal Model of Pulmonary Fibrosis

The amorphous SiO_2 nanoparticles (SNPs) (Code Name: SP1, purity: 99%, average diameter: 20±5nm, surface area: 640±30 m^2/g, hydroxyl concentration>45%) were provided by Zhejiang Hongsheng Nano-materials Technologies Co Ltd. Rats were instilled with intratrachealnano-SiO_2 particles suspension to establish the model of pulmonary fibrosis and the change of morphology and histopathology was investigated based on that model, as previously described by Zhang [8]. Briefly, rats were endotracheal instillation of 6.25 mg, 12.5 mg, 25 mg of nano-SiO_2 particles in 1 ml saline for 7d, 15d, 30d, 60d and 90d, while the rats of control group just endotracheal instillation of equivoluminal saline, then rats' lung tissue were removed. In this study, rats' lung tissue of nano-SiO_2 particles of 12.5 mg/mL on day 60 was chosen for further research.

2.2 Prediction of Target Genes Regulated by miR-702-3p

Target genes of miR-702-3p were predicted using the Target Scan (http://www.targetscan.org/). Gene set enrichment analysis of the predicted target genes was significantly predicted via the Gene Ontology (http://www.geneontology.org) and signaling pathway analysis of the target genes was filtrated through the database of KEGG pathway analysis (http://www.genome.jp/kegg/pathway.html). Meanwhile the genes may be involved in fibrosis were screened out and BMP-7 was selected for the further research.

2.3 Culture of A549 Cells and Dual-luciferase Reporter Gene Assay

Human type II alveolar epithelial carcinoma cell line (A549 cell) was purchased from the Cell Bank of Chinese Academy of Sciences (Shanghai, China). A549 cells (1.0×10^5 per well) were cultured in Dulbecco's modified Eagle's medium (DMEM; Gibco, Invitrogen, USA) containing 10% fetal bovine serum (FBS), 100 U/ml penicillin and streptomycin in 5% carbon dioxide atmosphere at 37°C.

The 3'-UTR of the rat BMP-7 gene was amplified by PCR using rat genomic DNA as a template. And the rno-miR-702-3p target site in the BMP-7 3'-UTR was mutated from GGTGGGC to CCACCCG by PCR mutagenesis. The PCR primers of wild type or mutant type of BMP-7 3'-UTR were showed as Table 1. The PCR fragment was inserted into the firefly luciferase reporter vector pmiR-RB-REPORT™ (RiboBio, Guangzhou, China) between the restrictive sites Xho I and Not I downstream from coding region of Renilla luciferase gene. All the cloning was confirmed by sequencing. A549 cells were plated at 3×10^3 cells/well in 96-well plates. After reaching 40-60%, cells were co-transfected with 100 ng/μL of the luciferase reporters vector containing the wild-type or mutant 3'-UTR and 50 nM of the miR-702-3p mimic or miR-702-3p Non-target Control (RiboBio, Guangzhou, China) with riboFECT™ CP Transfection Kit (RiboBio, Guangzhou, China). After 48 hour transfection, luciferase activities were measured in cell lysates using a Luc-Pair™ Duo-Luciferase Assay Kit 2.0 (Genecopoeia, USA). The Renilla luciferase signal was normalized to the firefly luciferase signal for each individual analysis.

2.4 Prediction of Target Genes Regulated by miR-702-3p

Total RNA of lung tissues stored at -80°C was extracted with TRIzol®Regent (Invitrogen, USA) according to the specification. The concentration and purity of the total RNA were determined with Nanodrop2000 Micro-spectrophotometer (Thermo Scientific, USA). The reverse transcription of miR-702-3p, BMP-7, TGF-β1, Smad3, E-cad, α-SMA and COL3 were directly performed with miRNA specific stem-loop primers or Oligo d (T) and the Reverse Transcription System (Promega, USA) according to the manufacturer's specification and the newly synthetic cDNA template was used for further research.

Real-time PCR amplification (20 μL reaction system) was performed under the condition of 95°C initial denaturation for 2 min, 95°C denaturation for 15 s, 60°C annealing for 60s on StepOne Plus Real-Time PCR System (Applied Biosystems, USA) with the GoTaq®qPCR Master Mix (Promega, USA) according to the manufacturer's specification and experiments were performed in triplicate following by 40 cycles. The mRNA sequences (Generay, Shanghai, China) and miRNA sequences (Ribobio, Guangzhou, China) used for detecting in PCR analysis were showed as Table 2. The data of PCR result was analyzed via the comparative threshold cycle (Ct) method. U6 snRNA and β-actin were chosen to be the endogenous control for miRNA and mRNA respectively. $2^{-\triangle\triangle Ct}$ denoted the fold change of target genes in the test group relatived to the control group ($\triangle CT=CT_{\text{target gene}} CT_{\text{endogenous control}}$, $\triangle\triangle Ct=\triangle Ct_{\text{test}}-\triangle Ct_{\text{control}}$).

Table 1: Primer for the Wild-type or Mutant 3'-UTR of Rat BMP-7

primer name	sequences	Total length of PCR product
BMP-7_WT_F	GGCGG**CTCGAG**AAGTACAGAAACATGGTGGTC	793bp
BMP-7_WT_R	AAT**GCGGCCGC**ACACGTCCCTCCCACCTTC	
BMP-7_MUT_F	ATAACCAG**CCACCCG**GCTTACGTCTGTGCTGAA	
BMP-7_MUT_R	ACGTAAGC**CGGGTGG**CTGGTTATGCCCTTCCTC	

Table 2: The Gene Primer Sequences Used for Detection in PCR Analysis

Gene	Primer sequence
miR-702-3p and U6	Forward and Reverse: confidentiality
BMP-7	Sense: 5' GAGGGCTGGTTGGTATTTGACAT 3'
	Antisense: 5' AGGGTTGCTTGTTCTGGGGTC 3'
TGF-β1	Sense: 5' CTAATGGTGGACCGCAACAAC 3'
	Antisense: 5' CACTGCTTCCCGAATGTCTGA 3'
E-cad	Sense: 5' GCATCCGACCTTGCTAACG 3'
	Antisense: 5' CATCTCCAGAGTCCAGCACAAT 3'
α-SMA	Sense: 5' TGAGGTCGGTGCCCGTATT 3'
	Antisense:5' CGTTGGTCTTGGGGTCTGT 3'
COL3	Sense: 5' TTTAATGGATAGGGACTTGTGTGAA 3'
	Antisense: 5' GAGAGAGAGAGAAGCTGAGGGTAGG 3'
β-actin	Sense: 5'CTGAACCCTAAGGCCAACCG3'
	Antisense: 5'GACCAGAGGCATACAGGGACAA3'

2.5 Western Blot Analysis

Total protein was extracted with the RIPA (SunShine Biotechnology, Nanjing, China) from lung tissues according to the manufacturer's instructions. The concentration of total proteins was measured with BCA Protein Assay Kit (Pierce, Thermo, USA). The proteins were diluted with RIPA buffer to the desired concentration and then mixed with an appropriate amount of loading buffer (Thermo, Pittsburgh, USA) with the proportion of 5:1. The protein mixture was boiled for 5 min at the temperature of 100°C. The protein mixture (20 μg each lane) were separated using 10% SDS-PAGE gel electrophoresis (Beyotime Biotechnology, China) at a constant voltage of 80 V for 2.5 h at 4°C and transferred to PVDF membranes (Millipore, USA) at a constant current of 250 mA for 1 h at 4°C. Then the PVDF membranes were blocked with 5% nonfat dry milk in TBST at room temperature for 1 h and incubated at 4°C overnight with the following primary antibodies: BMP-7, TGF-β1, Smad3, E-cad, α-SMA, COL3 and GAPDH (Abcom, USA). The membranes were incubated with 1:2000 dilution of goat anti-rabbit HRP-conjugated secondary antibody (Abgent, USA) for 1 h at room temperature. Finally, the membranes were incubated in ECL reagent (Pierce, Thermo, USA) for HRP detection and then exposed to autoradiography film (Bio-Rad, USA) for band visualization.

2.6 Statistical Analysis

In our study, the SPSS 21.0 software was used to evaluate statistical significance with the Student's t-test. All experimental results were expressed as means ± SD ($\bar{}\pm s$). The values of $p < 0.05$ were statistically significant.

3 Results

3.1 BMP-7 was a Target Gene of miR-702-3p

There were 136 target genes identified that may be regulated by miR-702-3p predicted with the database TargetScan.org and BMP-7 was filtrated to be the target genes of miR-702-3p through the analysis of Gene Ontology and KEGG.

With the assistance of target-scan database, the predicted binding site of miR-702-3p and BMP-7 indicated that BMP-7 is probably a target gene regulated by miR-702-3p. The correlation with between miR-702-3p and BMP-7 was further verified with the Dual-Luciferase Reporter Assay. Co-transfection of miR-702-3p and the portion of the human BMP-7 3'-UTR carried by luciferase reporter significantly reduced luciferase expression by approximately 2-fold when compared with the control group. Simultaneously, co-transfection of miR-702-3p and the mutated BMP-7 3'-UTR at the predicted miR-702-3p site carried by the luciferase reporter failed to diminish the luciferase activities when compared with its control group (Fig. 1). This data indicated that miR-702-3p directly regulates the post-transcriptional manner of BMP-7.

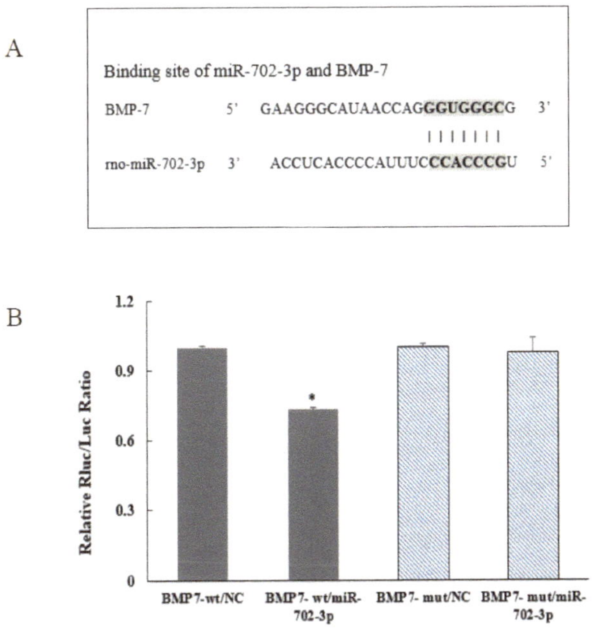

Figure 1: BMP-7 was a target gene of miR-702-3p. (A) Predicted binding site of miR-702-3p and BMP-7. MiR-702-3p seed sequence is in bold. The representation is limited to the region around the miR-702-3p complementary site. The number of interactional region ofBMP-7 3' UTR is 2090-2096. (B) Co-transfection of 50 nM miR-702-3p mimic or miR-702-3p NC and the portion of rat BMP-7 3'-UTR (WT or MUT) carried by the luciferase reporter operating with riboFECT™ CP Transfection Kit. After 48 h after transfection, the measured Renilla luciferase activity was normalized to the firefly luciferase signal for each individual analysis. *P<0.05.

3.2 Abnormal Expression of miR-702-3p and BMP-7 in Pulmonary Fibrosis Induced by Nano-SiO_2 Particles

After the experiments of Illumina Hiseq2000 high throughput sequencing and dual-luciferase reporter gene assay, the level expression of miR-702-3p and BMP-7 in pulmonary fibrosis induced by nano-SiO_2 particles was detected by Quantitative real-time PCR and western blot analysis. Results showed that the expression of miR-702-3p in the lung tissue of nano-SiO_2 particles group was up-regulated (fold changes > 2.0) when compared with the saline group (Fig. 2(A)), which was consistent with the result of Illumina Hiseq2000 high throughput sequencing. The expression of BMP-7 mRNA in the lung tissue of nano-SiO_2 particles group was down-regulated (fold changes < 0.5) when compared with the saline group (Fig. 2(A)) and the expression of BMP-7protein in the lung tissue of nano-SiO_2 particles group was significantly reduced (fold changes < 0.5) than the saline group ($P<0.05$) (Fig. 2(A) (B)).

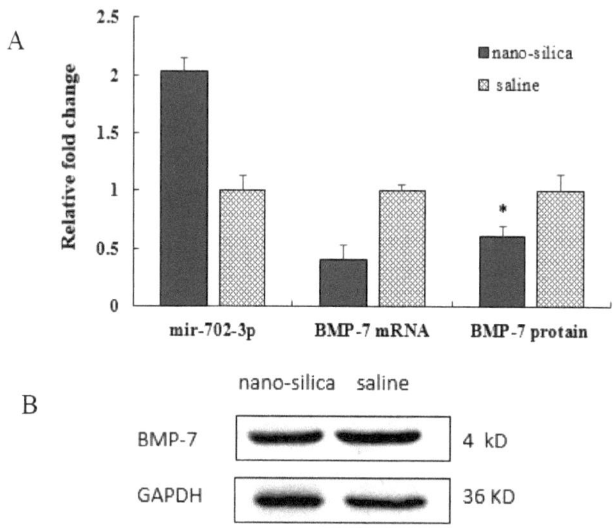

Figure 2: Abnormal expression of miR-702-3p and BMP-7 in pulmonary fibrosis induced by nano-SiO_2 particles. (A) The expression of miR-702-3p and BMP-7 in pulmonary fibrosis induced by nano-SiO_2 particles compared with the saline group. (B) Western Blot analysis for BMP-7. "Student's t-test" was used as statistical analysis. Values are expressed as mean + SD (n= 3). *$p < 0.05$: when compared with saline control group. BMP-7: bone morphogenetic protein 7.

3.3 EMT Contributed to Pulmonary Fibrosis Induced by Nano-SiO_2 Particles

Studies have demonstrated that EMT can contribute to bleomycin-induced lung fibrosis [33,35]. To analyze the correlation of EMT and pulmonary fibrosis induced by nano-SiO_2 particles, the expressions of epithelial and mesenchymal markers were determined by Quantitative real-time PCR and western blot analysis. Results of Quantitative real-time PCR showed that the expression level of TGF-β1 mRNA in the lung tissue of nano-SiO_2 particles group was up-regulated, Smad3 mRNA was up-regulated, E-cad mRNA was down-regulated, α-SMA mRNA was up-regulated and COL3 mRNA was up-regulated when compared with the saline group (Fig. 3(A)). Results from western blotting showed that the expression level of TGF-β1 protein in the lung tissue of nano-SiO_2 particles group was significantly up-regulated ($P<0.05$), Smad3 protein was significantly up-regulated ($P<0.05$), E-cad protein was significantly reduced ($P<0.05$), α-SMA protein was up-regulated and COL3 protein was significantly up-regulated ($P<0.05$), when compared with the saline group (Fig. 3(B) (C)). This data revealed that rat lung epithelial cells lost epithelial characteristics and acquired mesenchymal markers, which became fibroblast-like cells after the treatment of nano-SiO_2 particles. EMT was contributed to pulmonary fibrosis induced by nano-SiO_2 particles.

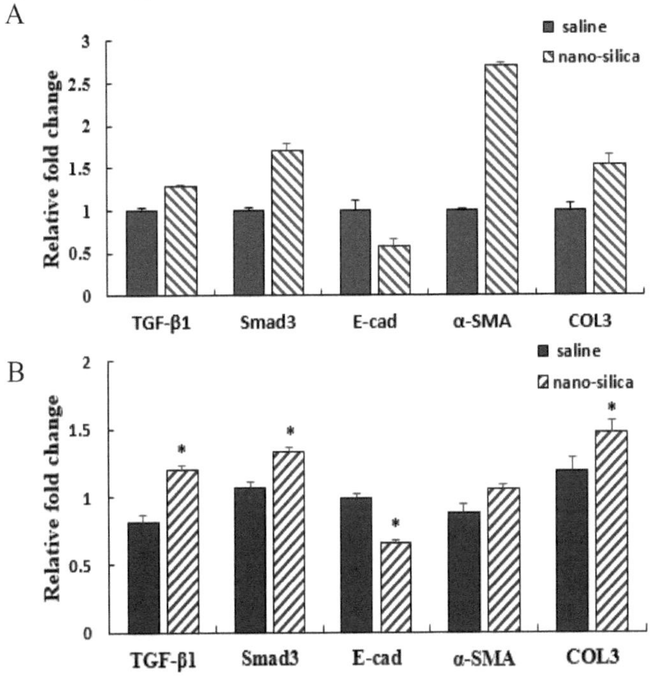

Figure 3: The expression of TGF-β1, Smad3, E-cad, α-SMA and COL3 in pulmonary fibrosis induced by nano-SiO_2 particles. (A) The expression of TGF-β1, Smad3, E-cad, α-SMA and COL3 mRNA. (B) The expression of TGF-β1, Smad3, E-cad, α-SMA and COL3 protein. (C) Western Blot analysis for TGF-β1, Smad3, E-cad, α-SMA and COL3. "Student's t-test" was used as statistical analysis. Values are expressed as mean + SD (n= 3). *$p < 0.05$: when compared with saline control group. BMP-7: bone morphogenetic protein 7.

C

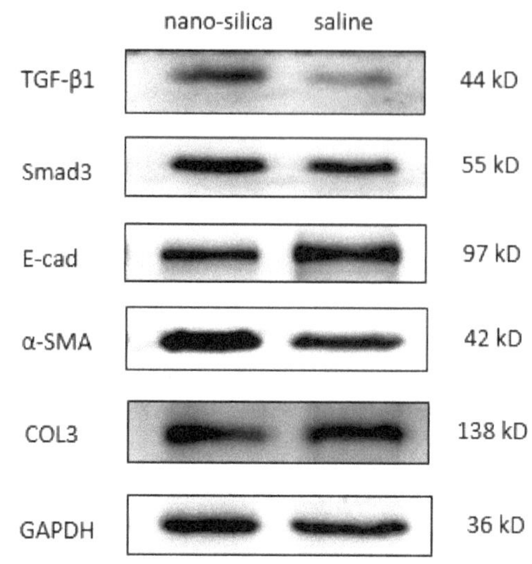

Figure 3: The expression of TGF-β1, Smad3, E-cad, α-SMA and COL3 in pulmonary fibrosis induced by nano-SiO_2 particles. (A) The expression of TGF-β1, Smad3, E-cad, α-SMA and COL3 mRNA. (B) The expression of TGF-β1, Smad3, E-cad, α-SMA and COL3 protein. (C) Western Blot analysis for TGF-β1, Smad3, E-cad, α-SMA and COL3. "Student's t-test" was used as statistical analysis. Values are expressed as mean + SD (n= 3). $^*p < 0.05$: when compared with saline control group. BMP-7: bone morphogenetic protein 7.

4 Discussion

Pneumoconiosis is a serious occupational disease characterized by fibrosing nodular lesions that eventually develop into progressive pulmonary fibrosis, which is induced by silicious dust, coal dust or other productive dusts [36]. Silicosis is an extremely serious pulmonary disease of pneumoconiosis, caused by the persistent inhalation of silica particles. Nano-SiO_2 particles have the same chemical composition with the micron-sized silica, but whether nano-SiO_2 particles can result in pulmonary fibrosis similarly still remains controversial. In recent years multitple studies have verified that nano-SiO_2 particles can induce cytotoxicity and lung injury of pulmonary inflammation and fibrosis in animal after intratracheal instillation of nano-SiO_2 particles [37,38]. Our previous studies have confirmed that rats had the symptom of pulmonary inflammation at protophase (7d, 15d and 30d) and pulmonary interstitial fibrosis atanaphase (60d and 90d) after treatment with nano-SiO_2 particles through endotracheal instillation [8]. Fibroblasts and myofibroblasts are effector cells to assemble abundance of ECM, which destroy the lungtissue structure and aggravate the fibrotic process [39]. Increasing evidence indicates that the mechanism of EMT may partially account for the origin of fibroblasts and myofibroblasts from alveolar epithelial cells [40]. TGF-β is a member of the superfamily of polypeptides which participate in cell proliferation, induction of apoptosis and the excessive accumulation of collagens. TGF-β is considered to be the dominating stimulus and TGF-β/Smad signal pathway is crucial for EMT in pulmonary fibrosis [41]. Additionally, BMP-7 has been identified as an antagonist of EMT [42]. BMP-7 can be in the preservation of the epithelial phenotype and reverse EMT induced by TGF-β in renal epithelial cells [18]. BMP-7 has been confirmed to antagonize fibrosis through restraining TGF-β/Smad signaling pathway [43]. The counterbalance of BMP-7 and TGF-β can control the process of EMT [44].

MiRNAs play a momentous regulatory role in the process of cell apoptosis, tissue inflammation and fibrosis, and miRNAs may regulate the development of pulmonary fibrosis. In our previous study, results from Illumina Hiseq2000 high throughput sequencing have shown that miR-702-3p is an up-regulated miRNA in fibrotic pulmonary tissue at 60 and 90 days [28]. With the help of Target-Scan database, the Gene Ontology and KEGG pathway analysis, BMP-7 was speculated to be one of the target genes regulated by miR-702-3p in participating in pulmonary fibrosis induced by nano-SiO_2 particles. In this study, the expression of miR-702-3p in the lung tissue of nano-SiO_2 particles group was up-regulated when compared with the saline group, which was consistent with the result from Illumina Hiseq2000 high throughput sequencing. The result indicated that miR-702-3p may promote the development process of pulmonary fibrosis induced by nano-SiO_2 particles. BMP-7, an antagonist of EMT [41], can reverse EMT induced by TGF-β in renal epithelial cells [18]. The result of this study showed that the expression of BMP-7 mRNA and protein in the lung tissue of nano-SiO_2 particles group were both down-regulated when compared with the saline group, which implied that BMP-7 may be an inhibited factor of pulmonary fibrosis induced by nano-SiO_2 particles. From all of the above data, together with the result of the Dual-Luciferase Reporter Assay, we deduced that BMP-7 was a target gene of miR-702-3p in regulating pulmonary fibrosis.

EMT, a trans-differentiation process from epithelial cells into mesenchymal cells, plays a crucial role in pulmonary fibrosis [10]. In the process of EMT, epithelial cells lose their markers, such as E-cad and cytokeratin, as well as mesenchymal cells get acquisition of markers, such as vimentin and a-smooth muscle actin (a-SMA) [11]. BMP-7, an antagonist of EMT, can inhibit tissue fibrosis through restraining TGF-β/Smad signaling pathway [42]. In this study, results showed that the expression level of TGF-β1 mRNA and protein in the lung tissue of nano-SiO_2 particles group were both up-regulated, Smad3 mRNA and protein were both up-regulated, E-cad mRNA and protein were both down-regulated, α-SMA mRNA and protein were both up-regulated and COL3 mRNA and protein were both up-regulated when compared with the saline group. The data revealed that in this process rat lung epithelial cells lost epithelial characteristics and acquired mesenchymal markers, which became fibroblast-like cells after the treatment of nano-SiO_2 particles. The above demonstrated that EMT contributed to pulmonary fibrosis induced by nano-SiO_2 particles.

5 Conclusions

MiR-702-3p may be a positive regulator in pulmonary fibrosis induced by nano-SiO_2 particles, and BMP-7 was a target gene of miR-702-3p in regulating pulmonary fibrosis. MiR-702-3p may directly weaken the post-transcriptional manner of BMP-7 to promote the process of EMT contributing to pulmonary fibrosis induced by nano-SiO_2 particles.

Acknowledgment: This work was supported by the National Natural Science Foundation of China (Grant No.81273046), the Fundamental Research Funds for the Central Universities, the Preventive Medicine Research Projects of Jiangsu Province (Grant No.Y2012039), and the Major State Basic Research Development Program of China (973Program, Grant No.2011CB933404).

References

[1] Fang S. C., Zhang H. T., Wang C. Y., & Zhang Y. M. Serum ca125 and nse: biomarkers of disease severity in patients with silicosis[J]. Clin. Chim. Acta, 2014, 433(7): 123-127.

[2] Mossman B. T., & Churg A. Mechanisms in the pathogenesis of asbestos and silicosis[J]. Am. J. Respir. & Crit. Care Med., 2012, 157(5 Pt 1): 1666-1680.

[3] Pollard K. M. Silica, silicosis, and autoimmunity[J]. Front. Immuno., 2016, 7(Suppl 5), doi: 10.3389/fimmu.2016.00097.

[4] Güngen A. C., Aydemir Y., Çoban H., Düzenli H., & Tasdemir C. Lung cancer in patients diagnosed with silicosis should be investigated[J]. Respir. Med. Case Rep., 2016, 18: 93-95.

[5] Coccini T., Barni S., Vaccarone R., Mustarelli P., Manzo L., & Roda E. Pulmonary toxicity of instilled cadmium-doped silica nanoparticles during acute and subacute stages in rats[J]. Histol. & Histopathol., 2013, 28(2): 195-209.

[6] Choi M., Cho W. S., Han B. S., Cho M., Kim S. Y., & Yi J. Y., et al. Transient pulmonary fibrogenic effect induced by intratracheal instillation of ultrafine amorphous silica in a/j mice[J]. Toxicol. Lett., 2008, 182(1-3): 97-101.

[7] Yang P., Zhang J., Y. Xia, Huang F., Xu Y. Y., Zheng Y. F., & Zhu X. Q. the pulmonary toxicity study of nano-silica particles on rats through dynamic inhalation[J]. Zhonghua lao dong wei sheng zhi ye bing za zhi, 2013, 31(7), 487-491.

[8] Zhang Y., Li W., Zheng Y., Wang X., Li G., & Yang H. dynamic changes of pathological morphology and ultrastructure of lung injury in rats induced by SiO_2 nanoparticles. Zhonghua lao dong wei sheng zhi ye bing za zhi, 2014, 32(7), 504-510.

[9] Gui T., Sun Y., Shimokado A., & Muragaki Y. The roles of mitogen-activated protein kinase pathways in tgf-β-induced epithelial-mesenchymal transition[J]. J. Signal Transduction, 2011, 2012(4): 289243-289243.

[10] Zavadil J., & Böttinger E. P. Tgf-beta and epithelial-to-mesenchymal transitions[J]. Oncogene, 2005, 24(37): 5764-5774.

[11] Chea S. W., & Lee K. B. Tgf-β mediated epithelial-mesenchymal transition in autosomal dominant polycystic kidney disease[J]. Yonsei Med. J., 2009, 50(1): 105-111.

[12] Câmara J., & Jarai G. Epithelial-mesenchymal transition in primary human bronchial epithelial cells is smad-dependent and enhanced by fibronectin and tnf-alpha[J]. Fibrogenesis & Tissue Repair, 2010, 3(1): 1-11.

[13] Chen K. C., Chen C. Y., Lin C. J., Yang T. Y., Chen T. H., & Wu L. C. Luteolin attenuates tgf-β1-induced epithelial–mesenchymal transition of lung cancer cells by interfering in the pi3k/akt–nf-κb–snail pathway[J]. Life Sci., 2013, 93(24): 924-933.
[14] Derynck R., Muthusamy B. P., & Saeteurn K. Y. Signaling pathway cooperation in tgf-β-induced epithelial–mesenchymal transition[J]. Curr. Opin. Cell. Biol., 2014, 31(31): 56-66.
[15] Han X., Yan S., Wei J. Z., Feng W., Liu X. W., & Meng Q. L. Critical role of mir-10b in transforming growth factor-β1-induced epithelial-mesenchymal transition in breast cancer[J]. Cancer Gene Ther., 2014, 21(2): 60-67.
[16] Kolosova I., Nethery D., & Kern J. A. Role of smad2/3 and p38 map kinase in tgf-β1-induced epithelial–mesenchymal transition of pulmonary epithelial cells[J]. J. Cell. Physiol., 2010, 226(5): 1248-1254.
[17] Kawashima T., Yamazaki R., Matsuzawa Y., Yamaura E., Takabatake M., Otake S., et al., Contrary effects of sphingosine-1-phosphate on expression of α-smooth muscle actin in transforming growth factor β1-stimulated lung fibroblasts[J]. Eur. J. Pharmacol., 2012, 696(1–3): 120-129.
[18] Xu Y., Wan J., Jiang D., & Wu X. Bmp-7 counteracts tgf-beta1-induced epithelial-to-mesenchymal transition in human renal proximal tubular epithelial cells[J]. J.Nephrol., 2009, 22(3): 403-410.
[19] Bartel D. P. Micrornas: target recognition and regulatory functions[J]. Cell, 2009, 136(2): 215-233.
[20] Inui M., Martello G., & Piccolo S. Microrna control of signal transduction[J]. Nat. Rev. Mol. Cell Biol., 2010, 11(4): 252-263.
[21] Wang H., Zhu L. J., Yang Y. C., Wang Z. X., & Wang R. Mir-224 promotes the chemoresistance of human lung adenocarcinoma cells to cisplatin via regulating g1/s transition and apoptosis by targeting p21waf1/cip1[J]. Brit. J. Cancer, 2014, 111(12): 339-354.
[22] Guo Z., Wu,R. Gong J., Zhu W., Li Y., & Wang Z. Altered microrna expression in inflamed and non-inflamed terminal ileal mucosa of adult patients with active crohn's disease[J]. J.Gastroenterol. & Hepatol., 2014, 30(1): 109-116.
[23] Weigel C., Schmezer P., Plass C., & Popanda O. Epigenetics in radiation-induced fibrosis[J]. Oncogene, 2014, 34(17): 2145-2155.
[24] Disayabutr S., Kim E. K., Cha S. I., Green G., Naikawadi R. P., & Jones K. D. Mir-34 mirnas regulate cellular senescence in type ii alveolar epithelial cells of patients with idiopathic pulmonary fibrosis. [J]. Plos One, 2016, 11(6): e0158367. doi: 10.1371/journal.pone.0158367.
[25] Zhou J., Tan Y., Chen S., Luo M., Pan L., & Liang W. Microrna-506 upregulation ameliorates pulmonary fibrosis by suppressing tgf-β1/smad3 pathway[J]. J. Biomaterials & Tissue Engineering, 2016. doi: http://dx.doi.org/10.1166/jbt.2016.1432.
[26] Kamikawaji K., Seki N., Watanabe M., Mataki H., Kumamoto T., Takagi K., et al., Regulation of loxl2 and serpinh1 by antitumor microrna-29a in lung cancer with idiopathic pulmonary fibrosis[J]. J. Human Genetics, 2016. doi: 10.1038/jhg.2016.99.
[27] Cui H., Banerjee S., Xie N., Ge J., Liu R. M., & Matalon S. Microrna-27a-3p is a negative regulator of lung fibrosis by targeting myofibroblast differentiation[J]. Am. J. Respir. Cell Mol. Biol., 2016.
[28] Dymacek J., Snydertalkington B. N., Porter D. W., Wolfarth M. G., Castranova V., Qian Y., et al., Mrna and mirna regulatory networks reflective of multi-walled carbon nanotube-induced lung inflammatory and fibrotic pathologies in mice[J]. Toxicol. Sci., 2015, 144(1): 51-64.
[29] Zhang Y J. Screening of Differentially Expressed MicroRNAs in Rats' Pulmonary Injury Induced by Nanosized SiO_2 [D]. nanjing: Southeast University, 2014.
[30] Kim B. M., & Choi M. Y. Non-canonical micrornas mir-320 and mir-702 promote proliferation in dgcr8-deficient embryonic stem cells[J]. Biochem. Biophys. Res. Commun., 2012, 426(2): 183-189.
[31] Zhang W. G., Chen L., Dong Q., He J., Zhao H. D., Li,F. L. & Li H. Mmu-mir-702 functions as an anti-apoptotic mirtron by mediating atf6 inhibition in mice[J]. Gene, 2013, 531(2): 235-242.

[32] Bin S., Li H., Xu Y., Qi S., Li T., Liu X., et al., Bmp−7 attenuates tgf−β1−induced fibroblast−like differentiation of rat dermal papilla cells[J]. Wound Repair & Regeneration, 2013, 21(21): 275-281.

[33] Yang G., Zhu Z., Wang Y., Gao A., Niu P., Chen L., et al., Bone morphogenetic protein 7 attenuates epithelial-mesenchymal transition induced by silica[J]. Hum. Exp. Toxicol., 2016, 35(1): 69-77.

[34] Pobezinsky L. A., Etzensperger R., Jeurling S. , Alag A., Kadakia T., & Mccaughtry T. M. Let-7 micrornas target the lineage-specific transcription factor plzf to regulate terminal nkt cell differentiation and effector function[J]. Nat. Immunol., 2015, 16(5): 517- 524.

[35] Tanjore H., Xu X. C., Polosukhin V. V., Degryse A. L., Li B. , & Han W. Contribution of epithelial-derived fibroblasts to bleomycin-induced lung fibrosis[J]. Am. J. Respir. Crit. Care Med., 2009, 180(7): 657-665.

[36] Ji X., Wu B., Jin K., Luo C., Han R., Chen M., et al., Muc5b promoter polymorphisms and risk of coal workers' pneumoconiosis in a chinese population[J]. Mol. Biol. Rep., 2014, 41(7): 4171-4176.

[37] Ojeajiménez I., Urbán P., Barahona F., Pedroni M., Capomaccio R. , & Ceccone G. A highly flexible platform for tuning surface properties of silica nanoparticles and monitoring their biological interaction[J]. ACS Appl. Mat. Interfaces, 2016. doi: 10.1021/acsami.5b11216.

[38] Delaval M., Boland S., Solhonne B., Nicola M. A., Mornet S., & Baezasquiban A., et al. Acute exposure to silica nanoparticles enhances mortality and increases lung permeability in a mouse model of pseudomonas aeruginosa pneumonia[J]. Part. Fibre. Toxicol., 2015, 12(1): 1-13.

[39] Song P., Zheng J. X., Xu J., Liu J. Z., Wu L. Y., & Liu C. β-catenin induces a549 alveolar epithelial cell mesenchymal transition during pulmonary fibrosis[J]. Mol. Med. Report., 2015, 11(4): 2703-2710.

[40] O'Beirne S. L., Walsh S. M., Fabre A., Reviriego C., Worrell J. C., & Counihan I. P. Cxcl9 regulates tgf-β1-induced epithelial to mesenchymal transition in human alveolar epithelial cells[J]. J. Immunol., 2015, 195(6): 2788.

[41] Wynn T. A., Integrating mechanisms of pulmonary fibrosis. J. Exp. Med., 208(7), 1339-1350.

[42] Zeisberg M., Hanai J., Sugimoto H., Mammoto T., Charytan D., Strutz F., et al., Bmp-7 counteracts tgf-beta1-induced epithelial-to-mesenchymal transition and reverses chronic renal injury. Nat. Med., 9(7), 964-968.

[43] Mitu G., & Hirschberg R. Bone morphogenetic protein-7 (bmp7) in chronic kidney disease. Front. Biosci., 13(13), 4726-4739.

[44] Kalluri R., Neilson E. G., Epithelial-mesenchymal transition and its implications for fibrosis[J]. China J. Mod. Med., 2003, 112(12): 1776–1784.

Jing Zhang*

Characteristics of Microbial Activities and Soil Nutrients in Soil Profiles of Tibet Alpine Wetlands

Abstract: To elucidate soil microbial activities and assess the soil quality, soil organic carbon (SOC), microbial biomass carbon (MBC), basal respiration (BR), potential respiration (PR), microbial quotient (qMB) and metabolic quotient (qCO_2) in soil profiles of marsh and meadow wetlands were investigated. The results showed that the SOC, MBC, BR, PR, and qMB decreased markedly with depth in 0~40 cm soil layer of marsh and meadow wetlands, while qCO_2 noticeably increased with depth. However, the SOC, MBC, BR, PR, qMB and qCO_2 remained more or less unchanged in 50~80 cm soil layer for both studied soils. Soil microbial activities in soil profiles of marsh wetland were lower than those in meadow wetland, especially in 0~40 cm layer. There were prominent positive relationships at $P<0.05$ in the soil profiles of marsh and meadow wetlands between MBC, BR, PR and qMB with soil organic carbon, total nitrogen, total phosphorus, available nitrogen, available phosphorus and available potassium. A significant negative relationship was found between the metabolic quotient with soil organic carbon, total nitrogen, total phosphorus and available nitrogen in both type soil profiles. With significant correlations between each microbial activity indicators identified. Soil microbial activities had strong correlations with soil measured chemical properties. The microbial activity indicators show that the soil quality in the soil profiles of marsh wetland is better than that in meadow wetland.

Keywords: tibet alpine wetland, microbial activity, soil organic carbon, basal respiration, metabolic quotient.

1 Introduction

According to the statistical data by the wetland resources investigation, the area of wetlands which are higher than 3000 m were 10.52 million hm^2, in which there were 6 million hm^2 in Tibet. Tibet alpine wetland is the most unique wetland in the world, besides the common function of wetland, due to the importance of its location, environment-specific; the Tibet alpine wetland has more special function in normal economic, social, protection function than any other wetland.

Soil microbe is an important part of the ecosystem, which is the most active component in soil. Soil microorganism is the power of soil substance transformation,

***Corresponding author: Jing Zhang,** Department of Resources and Environment, Tibet Agricultural and Animal Husbandry College, Linzhi, Xizang Province, 860000, China,E-mail: tonizhjing@hotmail.com and takes part in the circulation of soil carbon, nitrogen and other elements, which plays an important role in decomposition, transformation and circulation of organic matter and nutrients [1,2]. The change of soil micropopulation, especially microbial physiological and ecological factor can predict the change of soil [3,4]. To elucidate the soil microbial activities and assess the soil quality, soil organic carbon, microbial biomass carbon, basal respiration, potential respiration, microbial quotient and metabolic quotient in soil profiles of marsh and meadow wetlands were investigated in this paper.

2 Material and Method

2.1 Study Area

Study area was located in the junction of Shenzha Country, Nima Country and Bange Country, named Selincuo Marsh Nature Reserve (31°29′~31°47′N, 89°19′~89°56′E), which is the biggest in Tibet and the second salty lakes in China. The attitude of the lake is 4530 m. The average temperature is -0.4°C and the average temperature in July and January are separately 15.9°C and -17.9°C The average annual precipitation is 290.8 mm, mainly concentrated in 6-9 months, accounting for 92.2% of the total annual rainfall. The sunshine duration is 2 897.4 hours and the annual percentage of sunshine is 65%. The high wind days which the wind speed is higher than 17 m/s reach to 90.8 days. In the southwest area, there are quite a lot small rivers formed by nyenchen tanglha mountain glaciers, ice and snow which feed into Lake Namtso. There are plenty of moors along the river and the southwest of the lake. Because of the evaporation, moors along the river are most brackish water marshes. Marshy soil and meadow soil are the main soil types in the area.

2.2 Soil Sampling

Soil samples were collected in summer 2014 according to the field survey. Five soil profiles were selected in each type of the wetlands. Each of the soil profiles were divided into 8 layers, 10 cm each layer. In total 40 samples were collected. Fresh samples were divided into two parts, one was stored at 4°C and the other was air-dried and hand sieved through a 1 mm sieve to remove roots and other debris.

2.3 Method

1. *Soil nutrients:* Soil organic matter (SOC) was determined by the $K_2Cr_2O_7$–H_2SO_4 oxidation method. Total soil nitrogen (TN) was analyzed using the standard

Kjeldahl acid-digestion method. Soil available nitrogen (Av-N) was determined by the alkaline diffusion method and available phosphorus (Av-P) was determined by the Bray method. Soil phosphorus (TP) and potassium (TK) were measured using standard method of Soil agricultural chemistry analysis [5].

2. *Soil microbial biomass carbon (MBC):* Microbial biomass carbon in soil was estimated by fumigation extraction. Six portions equivalent to 25 g of dry weight soil were taken from each soil sample. Three portions were fumigated for 24 h at 25°C with $CHCl_3$ (ethanol-free). Following fumigant removal, the soil was treated with 100 ml of 0.5 M K_2SO_4 by horizontal shaking for 30 min at 200 rpm and then filtered. The other three non-fumigated portions were extracted simultaneously at the time fumigation commenced. Organic carbon in the extracts was measured using automatic carbon analyzer. The organic carbon difference (Ec) between the fumigated sample and non fumigated samples were calculated. Microbial biomass C was calculated as follows: MBC=Ec/0.45 [6].
3. *Soil basal respiration (BR) and respiration potential (PR):* Soil respiration was determined after trapping CO_2 in 0.1 mol dm^{-3} KOH solution and automated titration with 0.1 mol dm^{-3} HCl [7]. All samples were corrected for the CO_2 content of blanks. The factor 2.2 mg was used for the conversion of KOH to the µg CO_2. Basal respiration (BR) and potential respiration (PR) (addition of 2 g glucose per kilogram of soil) were measured after 24 h. Soil samples were incubated in the dark at 28°C and adjusted with distilled water to 50% the maximum water-holding capacity (MWHC) during the entire incubation. Results of the basal and potential soil respiration were expressed as µg CO_2-C per gram dry soil per hour. All analyses were realized in triplicate (respiration quadruplicate) and all results were calculated on soil dry matte.
4. *Soil microbial quotient (qMB) and the metabolic quotient (qCO_2):*

   qMB=MBC/SOC

   qCO_2=BR/MBC.

2.4 Data Analysis

One-way ANOVA analysis was used to test the difference of SOM contents and MBC contents. Pearson correlation coefficients were computed to analyze relationships between soil microbial activity and soil nutrients. Analyses were conducted using SPSS 10.0 statistical package and Origin 6.0 software package and differences considered significant if $p<0.05$.

3 Results

3.1 Distribution of SOC

Distribution of SOC content under different depths in two wetlands was shown in Fig. 1. From Fig. 1 we can see that with the increase of soil depth SOC content decreased and it showed a similar change in both wetlands. With the increase of soil depth in 0-40 cm, SOC contents in both wetlands noticeably decreased, but in 0-10 cm there were much higher SOC, 73.74 g kg^{-1} in marsh and 52.13 g kg^{-1} in meadow respectively. In the same soil depth, SOC content in marsh was much higher than that in meadow. There was a significant difference in SOC content between marsh and meadow in 0-50 cm ($P<0.05$), but there was no difference in 50-80 cm.

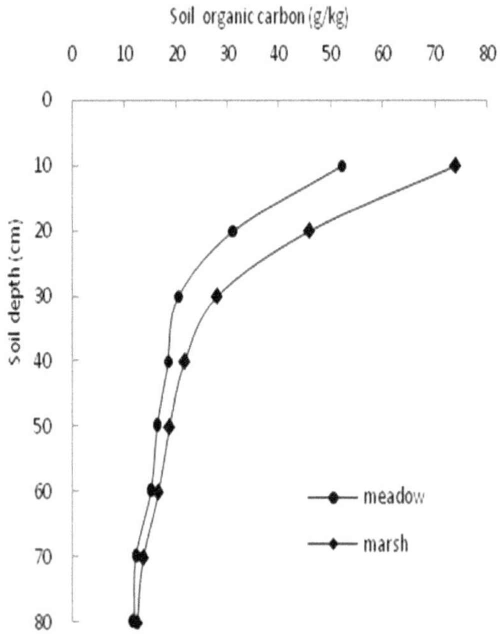

Figure 1: Distribution of SOC variation in soil profiles

3.2 Distribution of Soil Microbial Biomass Carbon

Distribution of soil microbial biomass carbon under different depths in two wetlands was shown in Fig. 2. Distribution of soil microbial biomass carbon showed a similar change with SOC. With the increase of soil depth in 0-40 cm, soil microbial biomass carbon decreased significantly, but in 0-10 cm there were much higher soil microbial biomass carbon, 2288.54 mg kg^{-1} in marsh and 1452.73 mg kg^{-1} in meadow respectively. In the same soil depth, soil microbial biomass carbon content in marsh was much higher than that in meadow. There was a significant difference in soil microbial biomass carbon content between marsh and meadow in 0-50 cm ($P<0.05$), but there was no difference in 50-80 cm.

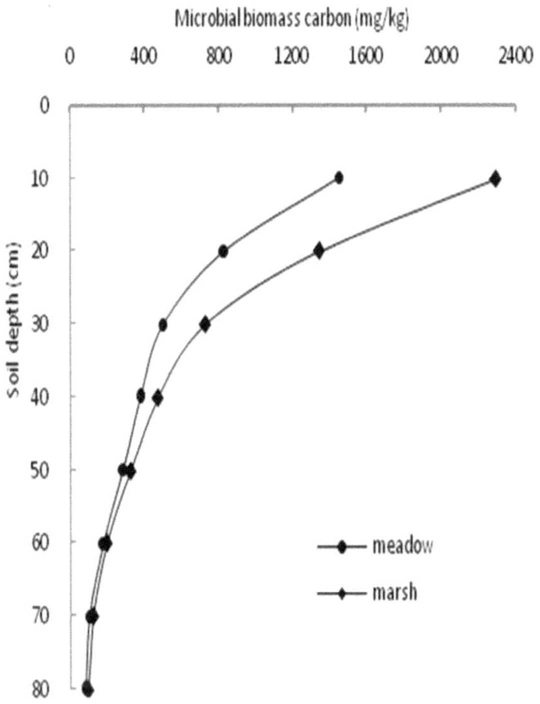

Figure 2: Distribution of MBC variation in soil profiles

3.3 Distribution of Soil Basal Respiration and Potential Respiration

From Fig. 3 we can see that with the increase of soil depth in 0-40 cm, soil basal respiration decreased significantly. Soil basal respiration in marsh of 0-10 cm was 5.41 mg kg^{-1} h^{-1} but that in meadow of 30-40 cm was only 1.43 mg kg^{-1} h^{-1}. Soil basal respiration of meadow in 0-40 cm was significantly lower than that in marsh ($P<0.01$), but there was no difference in 40-80 cm.

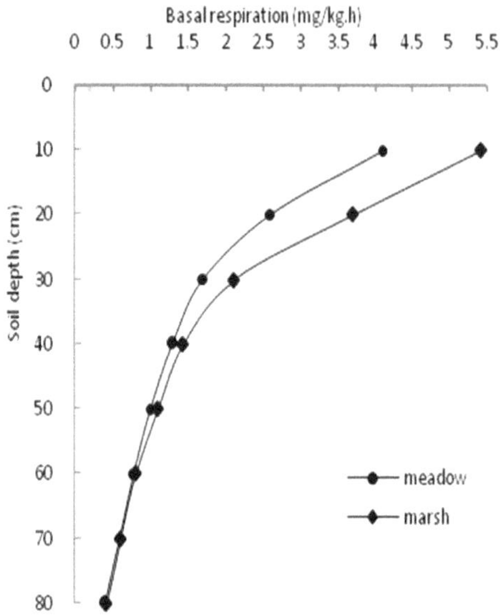

Figure 3 Distribution of BR variation in soil profiles

Soil potential respiration showed the same change as soil basal respiration from Fig. 4. With the increase of soil depth in 0-40 cm, soil basal respiration in both wetlands decreased significantly.

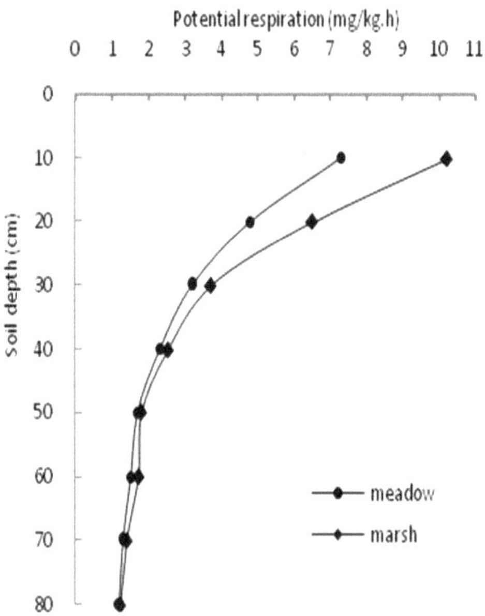

Figure 4: Distribution of PR variation in soil profiles

3.4 Distribution of Soil Basal Respiration and Potential Respiration

Soil microbial respiration quotient showed the same in both wetlands (Table 1). With the decrease of soil depth, the soil microbial respiration quotient decreased significantly in both wetland ($P<0.05$). Soil microbial respiration quotient in marsh wetland was much higher than that in meadow wetland in the 0-40 cm layer. Change of soil microbial metabolic quotient was shown in Table 1. Soil microbial metabolic quotient in each layer of marsh wetland was less than that of meadow wetland. With the increase of soil depth, soil microbial metabolic quotient increased in both wetlands. But soil microbial metabolic quotient in 60-70 cm of both wetlands was significantly higher than that in the 70-80 cm layer.

Table 1: Distribution of qmb And qco2 Variation in Soil Profiles

Depth (cm)	Marsh		Meadow	
	qMB (%)	qCO_2 ($mg \cdot kg^{-1} \cdot h^{-1}$)	qMB (%)	qCO_2 ($mg \cdot kg^{-1} \cdot h^{-1}$)
0-10	3.10a	2.36g	2.79a	2.82f
10-20	2.93a	2.75f	2.66a	3.13e
20-30	2.62b	2.89f	2.44b	3.39d
30-40	2.19c	3.03e	2.05c	3.44d
40-50	1.70d	3.45d	1.79d	3.49d
50-60	1.19e	4.07c	1.19e	4.33c
60-70	0.92f	4.86a	0.84f	5.68a
70-80	0.74f	4.33b	0.69g	4.89b

Note: Values are the means of three replicates. Means in a row followed by different letters differ significantly at $P<0.05$

3.5 Distribution of Soil Basal Respiration and Potential Respiration

Table 2 showed that there were close relations between soil microbial activity and main soil physicochemical properties. There was a significant positive correlation between soil microbial biomass carbon, soil basal respiration, soil potential respiration, soil microbial respiration quotient and main soil physicochemical properties ($P<0.01$) while there was a negative correlation between soil microbial metabolic quotient and main soil physicochemical properties. There was a significant correlation between soil microbial respiration quotient and soil total nitrogen, soil total phosphorus, soil available nitrogen and soil available phosphorus ($P<0.01$), which indicated that soil microbial activity was affected by soil physicochemical properties.

Table 2: Correlation Between Soil Microbial Activity Indicators and Chemical Properties

Item	Marsh					Meadow				
	MBC	BR	PR	qMB	qCO_2	MBC	BR	PR	qMB	qCO_2
SOC	0.981**	0.952**	0.985**	0.974**	-0.801**	0.980**	0.946**	0.972**	0.958**	-0.832**
TN	0.923**	0.943**	0.939**	0.897**	-0.673**	0.872**	0.912**	0.908**	0.876**	-0.618**
TP	0.937**	0.876**	0.872**	0.924**	-0.787**	0.907**	0.806**	0.845**	0.889**	-0.727**
Av-N	0.982**	0.945**	0.958**	0.953**	-0.756**	0.953**	0.908**	0.933**	0.901**	-0.705**
Av-P	0.915**	0.923**	0.856**	0.879**	-0.523**	0.899**	0.899**	0.809**	0.854**	-0.478*
Av-K	0.832**	0.798**	0.623**	0.711**	-0.478*	0.796**	0.732**	0.588**	0.658**	-0.312*

Note: Correlation is significant (two-tailed test) at * $P<0.05$ and ** $P<0.01$ n=40

4 Discussion and Conclusions

Wetlands, as an important component of global carbon cycle, account for about a third of the global carbon reserves, and have an important function of "carbon sink" [8,9]. Soil, the carbon and nitrogen repository of wetland, is the base and core of "carbon sink" in the wetland ecosystem [10-12]. Soil organic carbon of the two wetlands in this study were significantly different. Soil organic carbon of marsh wetland was much higher than that of meadow wetland in the 0-50 cm layer, which related to water condition. Soil moisture is not saturated in meadow wetland and soil aeration is in good condition, which is good for soil organic carbon decomposition. Soil microbial biomass carbon is an important index of soil microbial activity; it reflects the size of the soil microbial community size [13]. Change of soil microbial biomass carbon was similar with soil organic carbon. Correlation analysis showed that soil microbial biomass carbon had significant positive correlation with soil organic carbon, which indicated that soil microbial biomass carbon was restricted by soil organic carbon.

Soil microbial respiration quotient is an effective index to evaluate the quality of the soil organic carbon dynamics, to reflect the soil organic carbon conversion efficiency of microbial biomass carbon, soil minerals and soil carbon loss of organic carbon fixation [14,15]. Soil microbial respiration quotient of marsh wetland was higher than that of meadow wetland, which indicated that soil microbial utilization efficiency of marsh wetland is higher than that of meadow wetland.

Soil basal respiration is an indicator to measure soil microbial total activity and an important index to reflect the status of soil matrix [16]. From Table 2 we can see that there was a significant positive correlation between soil basal respiration and soil organic carbon. And there also was a significant linear correlation between soil basal respiration and soil microbial biomass carbon. This indicates that soil basal respiration was limited by soil microbial biomass carbon from the regression analysis.

Soil microbial metabolic quotient represents the ability that the soil microbial community can maintain the efficiency of soil matrix. The lower the soil microbial metabolic quotient, the higher the utilization efficiency of soil microbial on organic carbon. From Table 1 we can see that soil microbial metabolic quotient increased with soil depth, which indicated that the utilization efficiency of soil microbial on organic carbon in surface layer was higher than that in the sublayer. Generally in adverse situations, microorganisms must be diverted from the energy to sustain growth and reproduction to compensate for the additional energy that is required to be paid by stress, which cause the increase of soil microbial metabolic quotient. We can conclude that the environmental stress level of meadow wetland was greater than that of marsh wetland.

Soil microbial activity reflects the change of soil quality. In this study soil microbial activity of marsh wetland was higher than that of meadow wetland, which indicated that soil quality of marsh wetland was superior to that of meadow wetland.

Acknowledgment: The research was supported by National Natural Science Foundation of China (41261073) and Top-notch Young Teachers of Tibet Agricultural and Animal Husbandry College.

References

[1] He Ch.Q, Zhao K.Y, The accumulation, allocation and biological cycle of the nutrient elements in Carex lasiocarpa wetland[J], Acta Eco.Sin., 2001, 21, 2074-2080. [in Chinese].
[2] Xiong H.F, Liao Q.Zh, Wu Q.F, Wang Y.H, Distribution on Soil Nutrients and Correlation Analysis in Lake Liangzi Wetlands, Hubei Province[J], J. Lake Sci., 2005, 17, 93-96. [in Chinese].
[3] Yang G.Sh, Song Ch.Ch, Wan Zh.M, Song Y.Y, Microbial activity in soils of Calamagrostis angustifolia wetlands in the Sanjiang Plain[J], Acta Sci. Circumstant., 2010, 30, 1715-1721. [in Chinese].
[4] Liu Y.Y, Li F, Sun Q.Y, Xie Y.H, Review on the Study of Soil Microorganisms in Wetland Ecosystems[J], Chin. J. Appl. Environ. Biol., 2013, 19, 547-552. [in Chinese].
[5] Lu R.K, Soil agricultural chemistry analysis[M], Beijing: China's agricultural science and technology press, 1999, 106-132. [in Chinese].
[6] Hofman J., Bezchlebová J., Duek L., Dolezal L., Holoubek I., Andel P., et al., Novel approach to monitoring of the soil biological quality[J], Environ. Int., 2003, 28, 771-778.
[7] Wu J, Brookes P.C, The proportional mineralization of microbial biomass and organic matter caused by air-drying and rewetting of a grassland soil[J], Soil Biol. Biochem., 2005, 37, 507-515.
[8] Xiao Y, Huang Zh.G, Wu H.T, L(u) X.G, Carbon and nitrogen distributions and microbial characteristics in the soils of four types of wetlands in Sanjiang Plain Northeast China[J], Chin. J. Appl. Ecol., 2014, 25, 2847-2854. [in Chinese].
[9] Gao J.Q, Lei G.Ch, Li L, Lu C, Bai M.Y, The Distribution Characteristics of Soil Organic Carbon in Three Kinds of Wetland Soils in Zoig(e) Plateau[J], Wetland Sci., 2010, 8, 327-330. [in Chinese].
[10] Yang Y.X, Wang Sh.Y, He T.R, Distribution characteristics and seasonal dynamics of phosphorus and potassium in wetland ecosystem in the Sanjiang Plain[J], Chin. J. Appl. Ecol., 2001, 12, 522-526. [in Chinese].

[11] Gao J.Q, L(u) X.G, Vertical Variabil ity of Major Nutrition Elements in Carex Lasiocarpa Wetland Soil and Farmland Soil[J], Bull.Soil Water Conserv., 2002, 22, 32-34. [in Chinese].
[12] Bai J.H, Deng W, Zhang Y.X, Spatial Distribution of Soil Organic Matter and Nitrogen in Soil of Circular-Zonary Vegetation Areas in Wulanpao Wetland,Inner Mongolia[J], J. Lake Sci., 2002, 14,145-151. [in Chinese].
[13] Wang Ch.T, Long R.J, Wang Q.L, Jing Z.Ch, Shi J.J, Du Y.G, et al., Changes in Soil Organic Carbon and Microbial Biomass Carbon at Different Degradation Successional Stages of Alpine Meadows in the Headwater Region of Three Rivers in China[J], Chin. J. Appl. Environ. Biol., 2008, 14, 225-230. [in Chinese].
[14] Yang G.Sh, Song Ch.Ch, Song Y.Y, Hou C.C, Characteristics of microbial activities in soil profiles of Calamagrostis angustifolia wetlands in Sanjiang Plain[J], Acta Eco. Sin., 2010, 30, 6146-6153. [in Chinese].
[15] Zhang J, Ma L, Ding X.H, Chen X.R, Ma W, Seasonal dynamics of soil microbial biomass C and N in different habitats in Zhalong Wetland[J], Acta Eco. Sin., 2014, 34, 3712-3719. [in Chinese].
[16] Li G.H, Chen Q.F, Huang Y.M, An Sh.Sh, Zheng F.L, Chen L.D, Soil microbial biomass C, N, P and basal respiration in rhizosphere soil of typical plants on the Loess Plateau[J], Acta Eco. Sin., 2010, 30, 976-983. [in Chinese].

Peipei Wang, Lingling Zhao*

The Expression Change of GFAP and HMGB1 in Primary Cultured Astrocytes Exposed to Chlorpyrifos and Lipopolysaccharide

Abstract: *Background and Objective:* During the development of the nervous system, the impact of adverse environmental factors can cause serious consequences. Chemicals, toxic substances and infections are considered as common risk factors. In order to find the impact of environmental pollutants pesticides and infections on the nervous system, we attempt to investigate the effect of CPF (chlorpyrifos) and LPS (lipopolysaccharide) combined exposure on the release of GFAP/HMGB1 and activation in the primary cultured astrocyte of neonatal rat cerebral cortex. *Methods*: In this paper, trypsin digestion was used to isolate and purify astrocytes, and identify the cell based on immunofluorescence method; Astrocyte in experiment group was exposed to different concentrations of LPS and CPF. Astrocyte's activities were observed via CCK-8; Immunofluorescence and Western blot were used to measure the changes on the expression level of GFAP as well as HMGB1 in astrocyte. *Results*: Astrocyte's synaptic network was significantly increased at 24 h after being exposed to 25, 50 μMCPF combined with 1 μg/mL LPS; The expression level of glial fibrillary acidic protein (GFAP). High mobility group box 1(HMGB1) increased after astrocytes were exposed to 25, 50 μM CPF combined with 1 μg/mL LPS ($p<0.05$), The expression level of GFAP and HMGB1 decreased after being exposed to 100 μM CPF combined with 1 μg/mL LPS($p<0.05$). 4) *Conclusion*: 1. CPF and LPS can not only cause astrocyte activation but also increase the expression of GFAP and HMGB1. 2. The effect of CPF combined with LPS is more obvious than CPF or LPS respectively on astrocytes, which indicates that they have a synergistic effect.

Keywords: chlorpyrifos, lipopolysaccharide, astrocytes, GFAP, HMGB1.

1 Introduction

For more than a century, many researchers have mainly focused on early life as a source of adult disorder. As suggested by recent studies, the complex interactions of genes and environment may be the cause of Neurodevelopmental Disorders [1]. Pesticides are common environmental chemicals, and people are usually exposed

***Corresponding author: Lingling Zhao,** Pediatrics, The Third Xiangya Hospital of Central South University (CSU), Changsha, China, E-mail: llzhao2011@qq.com, fureya711@qq.com
Peipei Wang, Pediatrics, The Third Xiangya Hospital of Central South University (CSU), Changsha, China

to low doses of pesticides over a long term period. However, there are no obvious symptoms of pesticide poisoning. With in-depth studies on the low doses of pesticides, their harmful effects are increasingly recognized by the public [2]. During the development process, infection is a common cause of illness. Although current studies focus on single factors, the growth process is often the combined action by the environment in a variety of factors [3]. Recently, double-hit models are used to study chronic/tonic pain [4] as well as stress dysregulation [5]. To analyze whether the interaction between the infection and pesticides will significantly affect the nervous system, we designed an in vitro study to investigate the double-hit of CPF and LPS, and selected astrocytes as targets, GFAP and HMGB1 as an index, so as to provide new ideas to combat Neurodevelopmental Disorders.

2 Methods

2.1 Cell Culture

Primary astrocyte cultures were prepared from the cerebral cortices of 1 to 3-day-old SD rats. Furthermore, isolated cells were maintained in DMEM/F12 (HyClone), supplemented with 10% of fetal bovine serum in an atmosphere of 5% CO_2/humidified air (95%) at 37°C. After cells enrichment, Astrocyte was detached from the culture flask with 0.1% trypsin and 0.04% EDTA in Hank's balanced salt solution (Sigma-Aldrich). Notably, this was repeated three times. Astrocyte was plated onto poly-L-lysine-coated glass coverslips, and supplemented with culture medium. Experiments were carried out 2-3 days after cell plating.

2.2 Intervention Method

The DMEM/F12 culture mediums with different chlorpyrifos and lipopolysaccharide concentrations were utilized respectively to cultivate the prepared astrocyte for 12/24/48 hours. After that, PBS washed cells can be applied for the next processing.

2.3 Immunofluorescence

Isolated cells were characterized by the primary antibody against GFAP as well as its corresponding secondary fluorescent antibody, and by the blue fluorescent nuclear stain DAPI.

2.4 Western Blot

The whole astrocytes were prepared as described above. Meanwhile, the protein concentration was determined using a protein assay kit, and the separation method is SDS-PAGE. Then, the gel was electroblotted onto a nitrocellulose membrane. The membranes were probed with anti-GFAP (1:400 dilution), anti-HMGB1 (1:1000 dilution). Furthermore, filters were washed and incubated for 2 h at the room temperature.

3 Results

3.1 CPF and LPS Activated Astrocytes

Primary cultured neonatal rat's brain astrocytes exposed to LPS (1 μg/mL) combined with chlorpyrifos (50 μM) established cell model, DMSO group, PBS group, chlorpyrifos (50 μM) group, lipopolysaccharide (1 μg/mL) group in the control group. Moreover, GFAP expression was observed by immunofluorescence after 24 h. CyTM3- red: GFAP expression; DAPI- blue: nuclei. A: After the DMSO treated 24 h; B: PBS treated 24 h; C: CPF (50 μM) treated 24 h; D: LPS (1 μg/mL) treated 24 h; E: LPS (1 μg/mL) + CPF (50 μM) treated 24 h.

In this study, GFAP antibody was labeled with a red fluorescent dye CyTM3 that astrocytes cells were red. As shown by the results, both the star-shaped glial cells and synaptic contact increased after the exposure to LPS combined with low-dose chlorpyrifos for 24 h. These findings suggested that astrocytes were active (Figure 1).

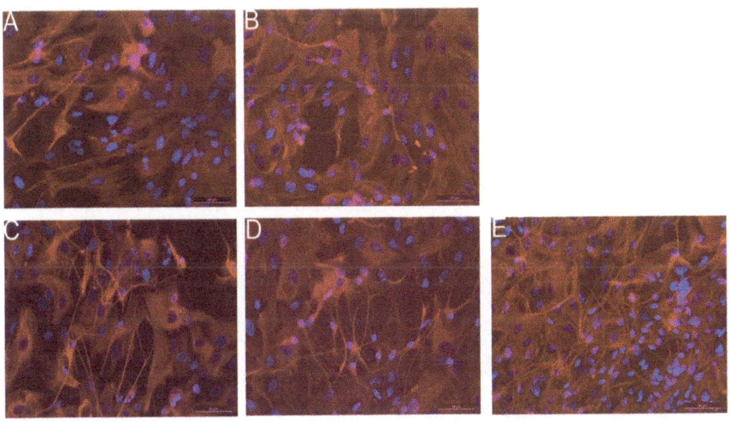

Figure 1: Expression of GFAP by astrocytes exposed to chlorpyrifos and lipopolysaccharide (immunofluorescence, × 400).

3.2 CPF and LPS Influences the Expression of GFAP

Primary cultured neonatal rat cortical astrocytes were exposed to chlorpyrifos (25 µM, 50 µM, 100 µM) and LPS (1 ug/mL); chlorpyrifos (25 µM, 50 µM, 100 µM) group, LPS (1 ug/mL) group, PBS (0.1%) groups, and DMSO (0.1%) in the control group. GFAP expression was observed by western blot after 12, 24, 48 h. A figure shown GFAP protein expression to detect changes at different time points, β-actin as internal control. B is a chart GFAP protein expression. *: P <0.05; x, group compared with the same time PBS and DMSO; a: compared with the same time CPF, b: compared with the same time LPS, compared with different time points, c: compare to 12 h, d: compare to 24 h, e: compare to 48 h.

The expression of GFAP was examined after astrocytes were simultaneously exposed to low-dose CPF and LPS via Western blot. As displayed by the results, the expression of GFAP in astrocytes exposure to chlorpyrifos (25, 50 µM) and LPS was gradually increased than the control group (p <0.05), while chlorpyrifos (100 µM) and LPS group was decreased (p <0.05) (Fig. 2).

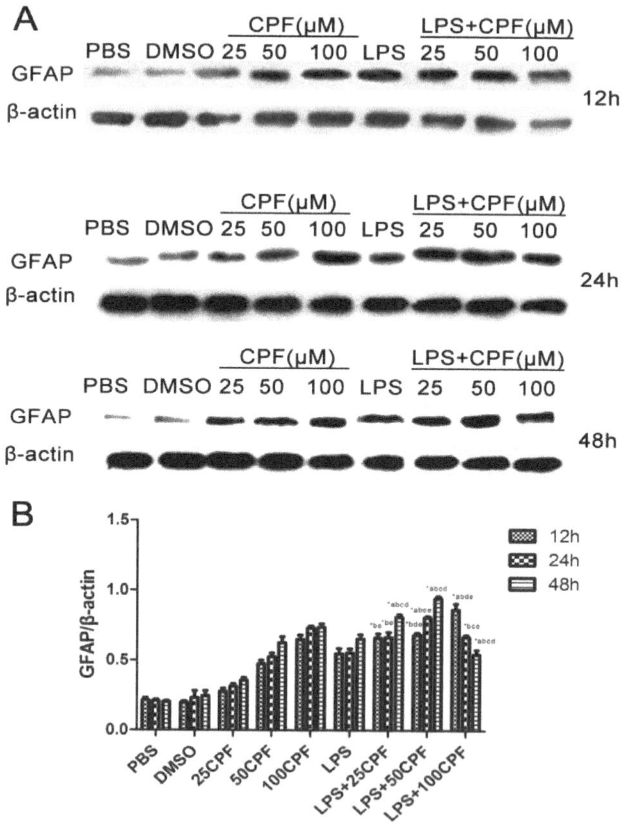

Figure 2: Expression of GFAP by astrocytes exposed to chlorpyrifos and lipopolysaccharide

3.3 CPF and LPS influences the expression of HMGB1

Primary cultured neonatal rat cortical astrocytes were exposed to chlorpyrifos (25 μM, 50 μM, 100 μM) and LPS (1 ug/mL); chlorpyrifos (25 μM, 50 μM, 100 μM) group, LPS (1 ug/mL) group, PBS (0.1%) groups, and DMSO (0.1%) in the control group. HMGB1 expression was observed by western blot after 12, 24, 48 h. A figure shown GFAP protein expression to detect changes at different time points, β-actin as internal control. B is a chart GFAP protein expression. *: $P < 0.05$; x, group compared with the same time PBS and DMSO; a: compared with the same time CPF, b: compared with the same time LPS, compared with different time points: c: compare to 12 h, d: compare to 24 h, e: compare to 48 h.

The expression of HMGB1 was examined after astrocytes were simultaneously exposed to low-dose CPF and LPS via Western blot. As shown by the results, the expression of GFAP in astrocytes exposure to chlorpyrifos(50 μM) and LPS was gradually increased than the control group ($p < 0.05$), while chlorpyrifos(100 μM) and LPS group was decreased ($p < 0.05$) (Fig. 3).

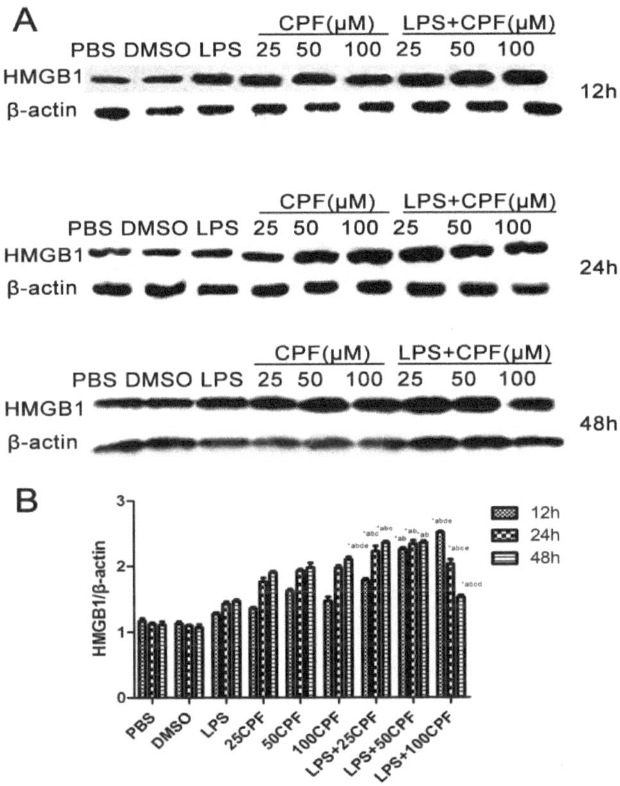

Figure 3: Expression of HMGB1 by astrocytes exposed to chlorpyrifos and lipopolysaccharide

4 Discussion

This experiment attempts to study a double-hit effect on the nervous system from an environmental chemical pollutant pesticides and infections, so as to explore the interaction between these two effects. Additionally, the primary cultured astrocytes were exposed to low-dose CPF and LPS *in vitro*. As shown by the results, low doses of CPF and LPS change its expression of GFAP and HMGB1. Moreover, the effect of their interaction is more obvious than their effect alone.

Generally speaking, the environmental pollution is considered as a risk factor for Neurodevelopmental Disorders [6]. According to the analysis of some scholars [7], the relationship between the environmental pollution and neurodevelopmental disorders and environmental attributable fraction (EAF) is 10%. People are often passively exposed to pesticides over a long period of time, and its health effects have attracted people's attention. Does the pesticide applied conform to "pesticide residue standards"? Is it really safe? A few studies have proposed that the "safe level" of toxic chemical residues is associated with neurodevelopmental disorders [8-9]. Organochlorine pesticides are persistent, and bioaccumulative environmental contaminants have potential neurotoxic effects. Moreover, the prenatal exposure to organochlorines is associated with impairment of neuropsychological development [10]. There is a growing emphasis on the impact of chlorpyrifos on children's neurodevelopment. As shown by animal or social investigation, the exposure to chlorpyrifos is closely related to children' nervous system damage [11], and it may cause neurodevelopmental impairment [10], mental retardation [12-14], ADHD [15], memory loss [16], anxiety and autism [17].

Previous studies from our group suggested that the sub-toxic doses of chlorpyrifos will induce brain inflammation [2], by HMGB1/TLR/NF-kB pathway in the brain amygdala regulated pro-inflammatory cytokines. Additionally, the HMGB1 could be the target of intervening and treating chlorpyrifos neurotoxicity [18].

Infections are a common reason attributed to sick children. As indicated by many studies, the intrauterine infection caused by fetal inflammatory response syndrome is the most common brain injury in preterm. The reason may be related to children's cognitive developmental disorders [19-21]. Inflammation is not only a potent inhibitor of neural development but also a high risk factor of cognitive impairment and depression [22-23]. A prospective study showed that 65% of very low birth weight children (gather of more Medical Center, a total of 6093 cases of newborns) had at least 1 infection. Compared with normal children, early childhood infections would significantly increase the risk of Neurodevelopmental Disorders [24].

The proliferation and differentiation of glial cells are later than neuron [25] because glial cells may be often affected by external factors during the maturity period. Many studies have suggested that the main target of chlorpyrifos is astrocytes [26-28]. Astrocytes refer to the important glial cells. Its features include the transfer guide neurons, glutamate and γ- amino butyric acid (GABA) metabolism, cell buffer of changes to the external environment, contact neurons through synapses. Also, it can accept a variety of materials through various receptor regulations, which results in nutritional factor [29]. GFAP is astrocytes' activation marker [30]. Moreover, a hallmark of gliotic reaction is the up-regulation of the astrocytic biomarker GFAP, which often precedes the anatomically apparent damage in the brain [31]. As shown by the experiments, CPF combined with LPS can give rise to the increase of astrocytes' expression of GFAP, suggesting that both CPF and LPS can cause astrocyte activation and nervous system damage. It is more obvious than each intervention alone, indicating that the two works together can jointly increase the activation of astrocytes and damage.

The important activity and function of HMGB1 lies in that it can be combined with various proteins. Also, it can mediate inflammation and gene transcription [32]. HMGB1 induces an inflammatory response via RAGE (receptor for advanced glycation end-products, RAGE) by glycosylation and Toll-like receptor family (toll-like family of receptors, TLRs) [33-34]. In this study, it was shown that CPF and LPS can cause the increase of HMGB1 expression in astrocytes, suggesting that CPF and LPS will work together to increase the nervous system injury. If the expression and secretion of HMGB1 is inhibited, can it alleviate the toxic effects of CPF and LPS on the nervous system? Can HMGB1 act as a common therapeutic target on these two kinds of injuries? This still needs further research.

5 Conclusions

From this study, it can be found that CPF and LPS will not only cause astrocyte activation but also increase the expression of GFAP and HMGB1. Additionally, the effect of CPF combined with LPS is more obvious than CPF or LPS respectively on astrocytes, suggesting that they pose a synergistic effect.

References

[1] Bale T L, Baram T Z, Brown A S, et al. Early Life Programming and Neurodevelopmental Disorders[J]. Biological Psychiatry, 2010, 68(4): 314-9.

[2] Zhang J, Dai H, Deng Y, et al. Neonatal chlorpyrifos exposure induces loss of dopaminergic neurons in young adult rats[J]. Toxicology, 2015, 336: 17-25.

[3] Gentile I, Altieri L, Lintas C, et al. Urinary polyomavirus infections in neurodevelopmental disorders[J]. Open Journal of Psychiatry, 2013, 03(2A): 18-25.

[4] Parent A J, Tétreault P, Roux M, et al. Descending nociceptive inhibition is modulated in a time-dependent manner in a double-hit model of chronic/tonic pain[J]. Neuroscience, 2016, 315(3): 70-78.

[5] Cotella E M, Durando P E, Suárez M M. A double-hit model of stress dysregulation in rats: implications for limbic corticosteroid receptors and anxious behavior under amitriptyline treatment.[J]. Stress-the International Journal on the Biology of Stress, 2014, 17(3): 235-46.

[6] Kishi R, Kobayashi S, Ikeno T, et al.Ten years of progress in the Hokkaido birth cohort study on environment and children's health: cohort profile--updated 2013. Environ Health Prev Med[J]. 2013, 18(6): 429-50.
[7] Philip J. Landrigan, Clyde B. Schechter, Jeffrey M. Lipton, et al. Environmental Pollutants and Disease in American Children: Estimates ofMorbidity, Mortality, and Costs for Lead Poisoning, Asthma, Cancer, andDevelopmental Disabilities[J].Environmental Health Perspectives.2002, 110(7): 721-728.
[8] Lisa M Chiodo, Sandra W Jacobson,Joseph L Jacobson. Neurodevelopmental effects of postnatal lead exposure at very low levels[J]. Neurotoxicology and Teratology. 2004, 26(3): 359–371.
[9] Mendola P, Selevan S G, Gutter S, et al. Environmental factors associated with a spectrum of neurodevelopmental deficits[J]. Mental retardation and developmental disabilities research reviews, 2002, 8(3): 188-197.
[10] Seyed Soheil Saeedi Saravi, Ahmad Reza Dehpour.Potential role of organochlorine pesticides in the pathogenesis ofneurodevelopmental, neurodegenerative, and neurobehavioraldisorders: A review[J]. Life Sciences.2016(145): 255-264.
[11] Rauh VA, Perera FP, Horton MK, et al. Brain anomalies in children exposed prenatally to a common organophosphate pesticide[J]. Proc Natl Acad Sci U S A, 2012, 109(20): 7871-7876.
[12] Bouchard MF, Chevrier J, Harley KG, et al. Prenatal exposure to organophosphate pesticides and IQ in 7-year-old children[J]. Environ Health Perspect, 2011, 119(8): 1189-1195.
[13] Rauh V, Arunajadai S, Horton M, et al. Seven-year neurodevelopmental scores and prenatal exposure to chlorpyrifos, a common agricultural pesticide[J]. Environ Health Perspect, 2011, 119(8): 1196-1201.
[14] Engel SM, Wetmur J, Chen J, et al. Prenatal exposure to organophosphates, paraoxonase 1, and cognitive development in childhood[J]. Environ Health Perspect, 2011, 119(8): 1182-1188.
[15] Fortenberry G Z, Meeker J D, Sanchez B N, et al. Urinary 3,5,6-trichloro-2- pyridinol (TCPY) in pregnant women from Mexico City: distribution, temporal variability, and relationship with child attention and hyperactivity[J]. Int J Hyg Environ Health, 2014, 217(2-3): 405-412.
[16] Horton M K, Kahn L G, Perera F, et al. Does the home environment and the sex of the child modify the adverse effects of prenatal exposure to chlorpyrifos on child working memory?[J]. Neurotoxicol Teratol, 2012, 34(5): 534-541.
[17] Aldridge JE, Levin ED, Seidler FJ, et al. Developmental exposure of rats to chlorpyrifos leads to behavioral alterations in adulthood, involving serotonergic mechanisms and resembling animal models of depression[J]. Environ Health Perspect, 2005, 113(5): 527-531.
[18] Tian J, Dai H, Deng Y, et al. The effect of HMGB1 on sub-toxic chlorpyrifos exposure-induced neuroinflammation in amygdala of neonatal rats[J]. Toxicology, 2015, 338: 95-103.
[19] Volpe Joseph J. Neonatal encephalopathy: an inadequate term for hypoxic-ischemic encephalopathy[J]. Ann Neurol. 2012, 72(2): 156-166.
[20] Adib-Conquy M, Cavaillon JM, Jacqmin S, et al. CD24-triggered caspase-dependent apoptosis via mitochondrialmembrane depolarization and reactive oxygen species pro-duction of human neutrophils is impaired in sepsis[J]. J Immunol, 2014, 192(5): 2449-2459.
[21] Kaur C, Rathnasamy G, Ling E A. Roles of activated microglia in hypoxia induced neuroin-flammation in the developing brain and the retina[J]. Journal of Neuroimmune Pharmacology, 2013, 8(1): 66-78.
[22] Zonis S, Ljubimov V A, Mahgerefteh M, et al. p21Cip restrains hippocampal neurogenesis and protects neuronal progenitors from apoptosis during acute systemic inflammation[J]. Hippocampus, 2013, 23(12): 1383-1394.
[23] Eisch A J, Petrik D. Depression and hippocampal neurogenesis: a road to remission?[J]. Science (New York, NY), 2012, 338(6103): 72.

[24] Stoll B J, Hansen N I, Adams-Chapman I, et al. Neurodevelopmental and growth impairment among extremely low-birth-weight infants with neonatal infection[J]. Jama, 2004, 292(19): 2357-2365.
[25] C. Guerri, J. Renau-Piqueras, Alcohol, astroglia, and brain development[J]. Mol. Neurobiol. 1997.15: 65–81.
[26] Krishnan K, Mitra N K, Yee L S, et al. A comparison of neurotoxicity in cerebellum produced by dermal application of chlorpyrifos in young and adult mice[J]. Journal of Neural Transmission, 2012, 119(3): 345-352.
[27] Salyha Y T. Chlorpyrifos leads to oxidative stress-induced death of hippocampal cells in vitro[J]. Neurophysiology, 2013, 45(3): 193-199.
[28] Lee J E, Lim M S, Park J H, et al. Nuclear NF-κB contributes to chlorpyrifos-induced apoptosis through p53 signaling in human neural precursor cells[J]. Neurotoxicology, 2014, 42: 58-70.
[29] Mercier F, Kitasako J T, Hatton G I. Anatomy of the brain neurogenic zones revisited: fractones and the fibroblast/macrophage network[J]. Journal of Comparative Neurology, 2002, 451(2): 170-188.
[30] Orre M, Kamphuis W, Osborn L M, et al. Isolation of glia from Alzheimer's mice reveals inflammation and dysfunction[J]. Neurobiology of aging, 2014, 35(12): 2746-2760.
[31] Ho G, Zhang C, Zhuo L. Non-invasive fluorescent imaging of gliosis in transgenic mice for profiling developmental neurotoxicity[J]. Toxicology and applied pharmacology, 2007, 221(1): 76-85.
[32] Yang H, Tracey K J. Targeting HMGB1 in inflammation[J]. Biochimica et Biophysica Acta (BBA)-Gene Regulatory Mechanisms, 2010, 1799(1): 149-156.
[33] Huebener P, Pradere J P, Hernandez C, et al. The HMGB1/RAGE axis triggers neutrophil-mediated injury amplification following necrosis[J]. The Journal of clinical investigation, 2015, 125(2): 539.
[34] Wang X, Sun R, Wei H, et al. High-mobility group box 1 (HMGB1)-Toll-like receptor (TLR) 4-interleukin (IL)-23-IL-17A axis in drug-induced damage-associated lethal hepatitis: Interaction of γδ T cells with macrophages[J]. Hepatology, 2013, 57(1): 373-384.

Hong Cao*, Yuansong Zhang, Jianqing Zai, Honggang Lai, Yunsheng Jiang, Huan Xiao, Yan Han

Optimization of Shelf Life of Fresh/Chilled Chicken Using Response Surface Method

Abstract: To prolong the shelf life of fresh or chilled chicken, the Oxford Cup Method was used to measure the inhibition effect of the preservatives, and selected out the appropriate concentration area of tea polyphenols is 0.08%-0.03%, Nisin is 0.03%-0.05%, and chitosan is 0.36%-0.60%, which had an obvious inhibition on spoilage bacteria, isolated from fresh or chilled chicken meat. According to the single factor experiment, the method of Box-Behnken experiment to design was used and response surface method optimization to study the effect of preservation of tea polyphenols, Nisin and chitosan on chicken breast. The result shows that when multiple preservative solutions is made with 0.03% tea polyphenols, 0.04% Nisin and 0.52% chitosan, the minimum value of TVB-N is 9.28009 mg/100 g. After proof test, the TVB-N is 9.32 mg/100 g which meets the test result, and explain that after optimized the regression model can be used. After immersed in the multiple preservative solutions, the shelf life of the chicken breast can be prolonged to 10 d at 4°C.

Keywords: fresh or chilled chicken, shelf life, Oxford Cup Method, spoilage bacteria, response surface method optimization, biopreservative, tea polyphenols, Nisin, chitosan.

1 Introduction

Fresh or chilled chicken is a meat product that after closer inspection drops in temperature to 0~4°C after 1 h, and is proceeded to be transported and sold [1]. Traditionally, people consume the chicken immediately. But recently the consumption pattern has changed, with people beginning to choose to buy fresh or chilled chicken. But after slaughter, the quality of the chicken is quickly reduced during a number of processes such as: proceeding, transporting, storing and selling, limiting the shelf life of chicken meat. Therefore, it is necessary to develop and study the technology of fresh or chilled chicken.

***Corresponding author: Hong Cao,** Jiangsu Province Lixiahe area agricultural research institute, Yangzhou, Jiangsu, China, E-mail: ch88188@163.com
Yuansong Zhang, Honggang Lai, Yunsheng Jiang, Yangzhou university tour and culinary art academy, Yangzhou, Jiangsu, China
Jianqing Zai, Huan Xiao, Yan Han, Jiangsu Province Lixiahe area agricultural research institute, Yangzhou, Jiangsu, China

2 Material and Method

2.1 Material and Equipment

1) Material and equipment:

a) Material and reagent: Local chicken: from Yangzhou area, bought in market, the quarantine of which had been qualified [3].

b) Biopreservative: Tea polyphenols, Chitosan, Nisin, food grade, served by Yangzhou KeDa Company.

c) Equipment: Biological microscopes, XS-18, Jiangnan optical instrument factory; PH meter, pHS-3C, Guangzhou Haosai Electronic Instruments factory; Multifunction Vacuum Packing Machine, DQB-36, Shanghai Qingpa Food Package Machinery company; Texture Analyzer, TMS-pro, Germany Food Technology Company; Automatic Colorimeter, SC-80C,Beijing Kangguang Optical Instrument Company; Refrigerator, BCD-206TX, Qingdao Haier Co.,Ltd. Oxford cup (internal diameter 6.0 mm±0.1 mm, high 10 mm±0.1 mm).

2) Method:

a) Destructive Test: Bled live chicken, dehaired, and discarded intestines. With the sterile water, washed the chicken body for 1min, and drained the water, put into the PE bag, stored at 4°C, for 10 d, the meat had already gone bad.

b) Isolation and identification of spoilage bacteria: Isolated the spoilage bacteria three times, vaccinate medium slant, and make a series of physiological and biochemical tests [4]. After identification, J1 is *Pseudomonas alcaligenes*, R1 is *Lactobacillaceae sharpeae*, C1 is *Citrobacter freundii*, and S1 is *B. thermosphacta*.

c) Oxford cup method: Put 15 mL Selective solid culture medium into each glass garden. Until solidified, drew 100 μL bacterial suspension on the medium and coat. Each group had three repetitions. Put six oxford cups in each glass garden, added 200 μL different concentrations of preservatives into the cup in turn. in addition the blank control was added without preservatives. Put glass gardens into Incubator according to the conditions stipulated for 24 h [5]. Measured the inhibition zone, calculated average values, and judged the effect of inhibition.

d) Three factors and five levels design of eesponse surface: According to the result of pre-experiment, tea polyphenols, chitosan, and Nisin, have obvious inhibition on the spoilage bacteria of fresh or chilled chicken. In this experiment set three factors as independent variables, and TVB-N as index. The factors and level of design of response surface shown as follows in Table 1.

3) TVB-N determination of fresh or chilled chicken: TVB-N determination: cut meat sample into pieces, weighed about 10 g, put into erlenmeyer flask, and then added 100 mL distilled water, shook, faltered after immersed for 30 min, measured the filtered liquor by the semi micro kjeldahl method [6].

Table 1: Three Factors and Five Levels Design of Eesponse Surface

Independent variables	Code	Level		
		-1	0	1
tea polyphenols	A	0.018	0.024	0. 3
Nisin	B	0.03	0.04	0.5
chitosan	C	0.36	0.48	0.06

3 Results

3.1 Identification result of Spoilage Bacteria

Results of physiological and biochemical identification as shown in Table 2.

Table 2: Results of physiological and biochemical identification

Biochemical Identification	R1	J1	C1	S1
Grow and Restore in 0.05% Potassium Tellurite	ND	ND	ND	+
Glucose	+	+	+	ND
Catalase	ND	ND	+	+
V-P	ND	ND	+	+
Arginine Digydrolase	ND	-	-	ND
Dioxygenase	-	+	-	ND
Hippurate Hydrolysis	ND	ND	ND	-
Gelatin Liquefaction	ND	+	-	ND
Fructose	+	-	ND	ND
Oxidase	-	+	-	ND
Rhamnose	+	ND	ND	-

"+"means positive, "-"means negative, "ND" means not detect.

According to the *common bacterial identification book*, as results of physiological and biochemical identification shows, combined with the biochemical reactions, initially identified J1 as *Pseudomonas alcaligenes*, R1 as *Lactobacillaceae sharpeae*, C1 as *Citrobacter freundii*, and S1 as *B. thermosphacta*.

3.2 Single-Factor Experiment

1) Inhabition of nisin on different isolation bacterial strain:

As Fig1 follows, Nisin has no obvious inhibition effect on *Pseudomonas alcaligenes* and *Citrobacter freundii*, because that Nisin has no inhibition on gram negative bacteria [7]. When the concentration is 0.02%, the inhibition zone of *B. thermosphacta* is 10.00 mm; when the concentration is 0.01%, the zone of *Lactobacillaceae sharpeae* is 14 mm. Therefore we can choose the better inhibition concentration which is 0.03%-0.05%.

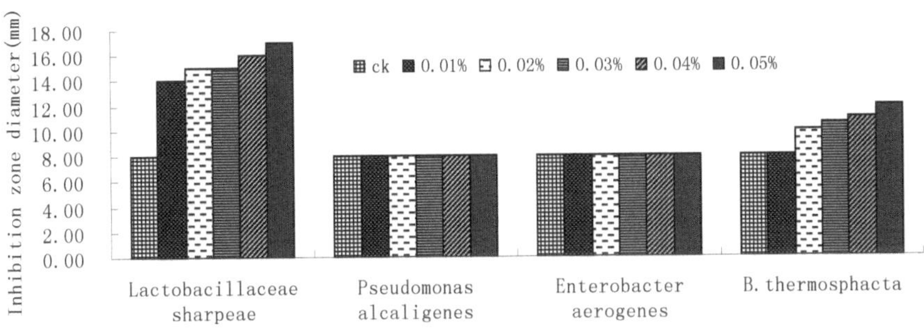

Figure 1: Inhibition of Nisin on the different isolation bacterial strain

2) Inhabition of chitosan on different isolation bacterial strain:

As Fig. 2 shows, chitosan had hardly any inhibition effect on *Lactobacillaceae sharpeae*, which related to the acidic solution chitosan dissolved [8]. Because *Lactobacillaceae* can grow up in the acidic circumstance, therefore it has no obvious effect on *Lactobacillaceae*. But chitosan has a strong inhibitory effect on *Pseudomonas alcaligenes*, *Citrobacter freundii* and *B. thermosphacta*, so we can choose the better concentration which is 0.36%-0.60%.

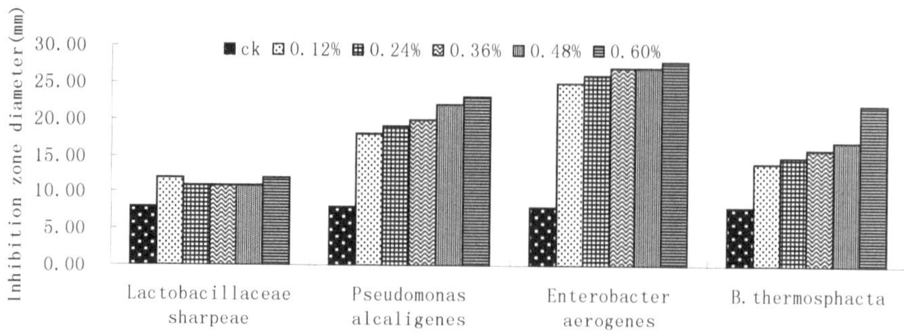

Figure 2: Inhibition of Chitosan on the different isolation bacterial strain

3) Inhabition of tea polyphenols on different isolation bacterial strain:
As Fig. 3 shows, tea polyphenols has obvious inhibition on *Pseudomonas alcaligenes*, *Enterobacter aerogenes* and *B. thermosphacta*, and hardly on *Lactobacillaceae sharpeae*. When the concentration is 0.006%, the inhibition zone of *Pseudomonas alcaligenes*, *Enterobacter aerogenes* and *B. thermosphacta* are respectively 15 mm, 14 mm and 13 mm; when the concentration is 0.006%, the inhibition zone of *Pseudomonas alcaligenes*, *Enterobacter aerogenes* and *B. thermosphacta* are respectively 20 mm, 19 mm and 18 mm. so we can choose the better concentration which is 0.018%-0.03%.

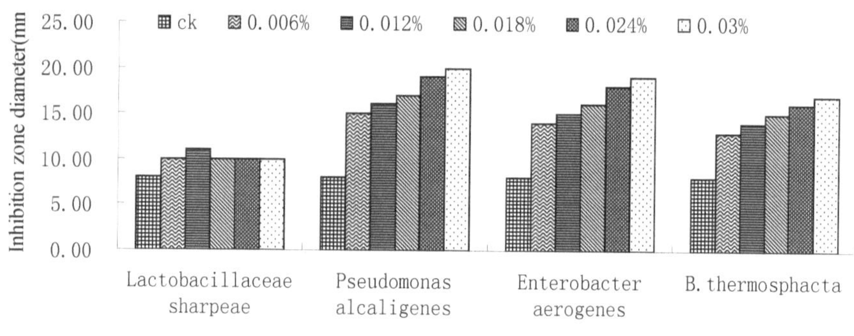

Figure 3: Inhibition of tea polyphenols on the different isolation bacterial strain

4) Result analysis of eesponse surface experiment:
With the software design expert 8.0.6, the method Box –Behnken was chosen to design the experiment [9]. On the basis of single factor experiments which chose tea polyphenols, chitosan and Nisin as factors, response surface methodology was employed to optimize the above three conditions as shown in Table 3.

Table 3: Three Factors and Five Levels Design of Eesponse Surface

Group	A%	B%	C%	TVB-N Actual (mg/100 g)	TVB-N predicted (mg/100 g)
1	0.024	0.03	0.36	14.65	14.65
2	0.018	0.04	0.6	12.96	13.04
3	0.024	0.05	0.6	9.98	10.04
4	0.024	0.04	0.48	10.32	10.25
5	0.03	0.04	0.6	9.72	9.75
6	0.024	0.03	0.6	12.61	12.59
7	0.024	0.04	0.48	10.17	10.25
8	0.018	0.03	0.48	14.52	14.66
9	0.018	0.05	0.48	12.82	12.87
10	0.018	0.04	0.36	15.3	15.02
11	0.03	0.05	0.48	9.57	9.61
12	0.024	0.05	0.36	12.85	12.82
13	0.03	0.04	0.36	12.53	12.62
14	0.024	0.04	0.48	10.31	10.25
15	0.024	0.04	0.48	10.18	10.25
16	0.03	0.03	0.48	12.13	12.22
17	0.024	0.04	0.48	10.33	10.25

5) Result analysis:
Analysis of the results of the experiment, the results of the analysis are shown in Table 4. The quadratic multiterm regression equation is

$$\textbf{TVB-N}=10.26-1.46*A-1.09*B-1.26*C-0.22*A*B-0.12*A*C-0.21*B*C+1.05*A^2+0.95*B^2+1.31*C^2$$

According to Table 3, model P<0.0001, which indicts that the regression model employed in the experiment is extremely significant, and has the better lack of fit.

Lack of fit is the probability of the equality between actual and predict. lack of fit P=0.3513>0.05, not significant, which indicts that in regression area the model is not lack of fit, and the residual error is random [10]. R2 of model is 0.9979, which indicts that degree of fitting of model is very good. Adjusted determination coefficient is 0.9979, which indicts that 99.79% of variability of data can be explained, but only 0.21% of variability can not be explained in the model. CV reflects confidence interval of model. CV% =0.74%. Adeq precision=85.403>4, which indicts that the experiment has little error, reliable data and high precision. So, in the model we can select the best proportion of compound natural preservative [11].

Table 4: The two response model of TVB-N analysis of variance

Source	Sum of squares	Freedom	Mean Square	F value	Prob>F
Model	56.9610	9	6.32901	829.5678	<0.0001**
A	16.9653	1	16.9653	2223.709	<0.0001**
B	9.43951	1	9.43951	1237.273	<0.0001**
C	12.6504	1	12.6504	1658.143	<0.0001**
AB	0.1849	1	0.1849	24.23556	0.0017**
AC	0.05522	1	0.05522	7.238554	0.0311*
BC	0.172225	1	0.17222	22.5742	0.0021**
A^2	4.65537	1	4.65537	610.1984	<0.0001**
B^2	3.77205	1	3.77205	494.4174	<0.0001**
C^2	7.26987	1	7.26987	952.891	<0.0001**
Residul Error	0.05340	7	0.00762		
Lack of Fit	0.02792	3	0.00930	1.461277	0.3513
Pure Error	0.02548	4	0.00637		
Total	57.0144	16			

To study the interaction effect of factors, software to draw response surface curve and contour line was used.

From the response surface curve and contour line, we can observe the significance level, circular means no significance between two factors, and ellipse is the opposite. When the concentration of tea polyphenolson and Nisin is increasing, TVB-N of fresh or chilled chicken is declining.

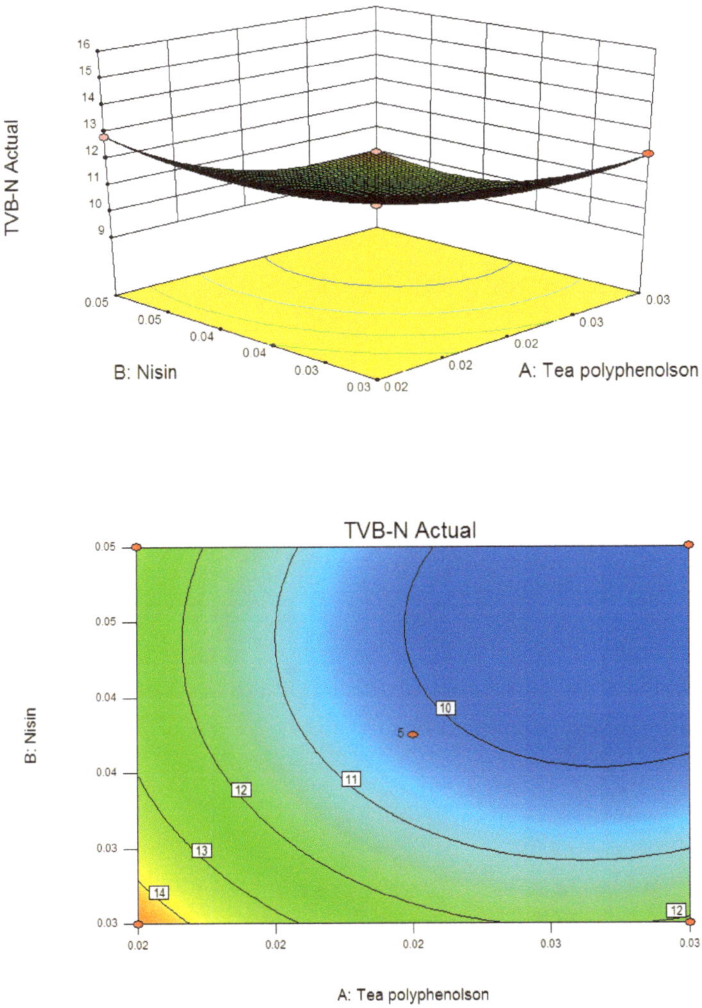

Figure 4: Inhibition of tea polyphenols on the different isolation bacterial strain

As shown in Fig. 5, the concentration of chitosan and Nisin has an interactive effect on TVB-N. In the experiment, when the concentration of chitosan and Nisin is increasing, TVB-N is firstly increasing, and then declining.

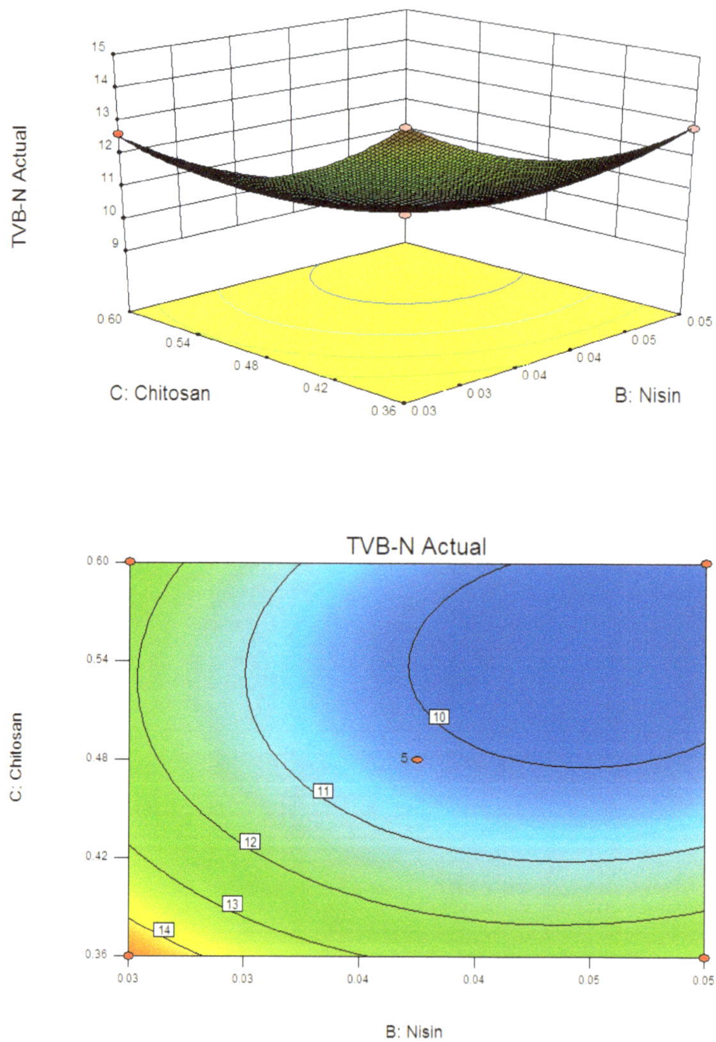

Figure 5: Response surface curve and contour line of effect of chitosan and Nisin on TVB-N of fresh or chilled chicken

As shown in Fig. 6, we can observe that the concentration of tea polyphenolson and chitosan has an interactive effect on TVB-N. In the experiment, when the concentration of tea polyphenolson and chitosan is increasing, TVB-N is firstly increasing, and then declining.

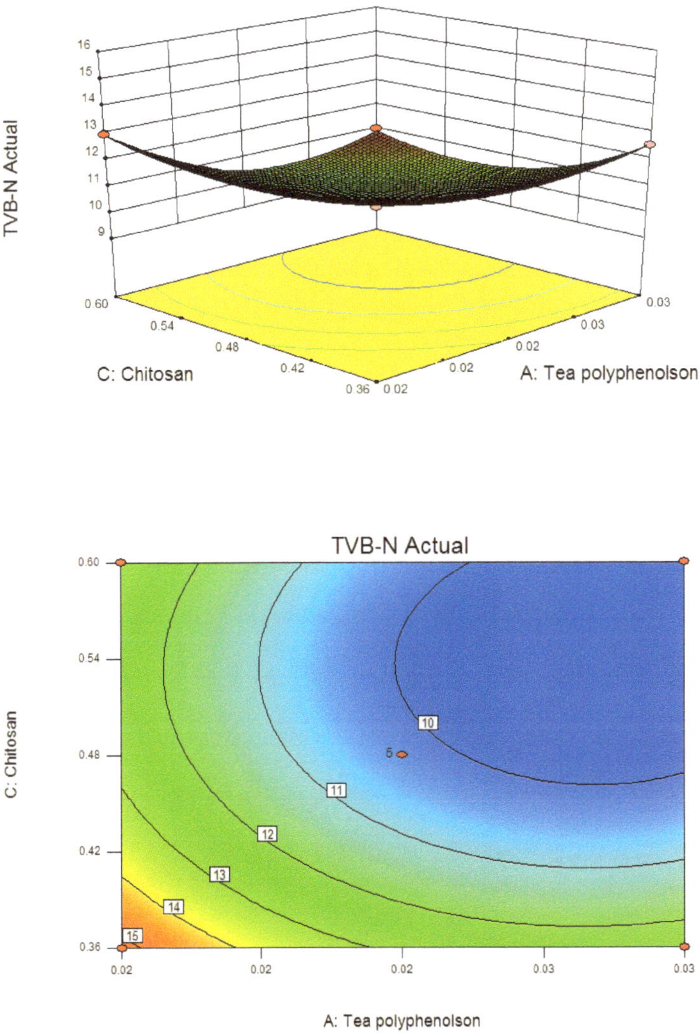

Figure 6: Response surface curve contour line of effect of tea polyphenolson and chitosan on TVB-N of fresh or chilled chicken

4 Conclusions

Isolated four chains of spoilage bacteria, identified J1 as *Pseudomonas alcaligenes*, R1 as *Lactobacillaceae sharpeae*, C1 as *Enterobacter aerogenes*, and S1 as *B.thermosphacta*. Used the Oxford Cup Method to measure the inhibition effect of the preservatives, and chose the appropriate concentration of tea polyphenolson is 0.08%-0.03%, Nisin is 0.03%-0.05%, and chitosan is 0.36%-0.60%, which have the better inhibition on spoilage bacteria. Then used the method Box–Behnken to design the experiment, set up 17 experimental groups to optimize the above three conditions. When the concentration of tea polyphenolson is 0.03%, Nisin is 0.04% and chitosan is 0.52%, the smallest value of TVB-N is 9.28009 mg/100 g. After the proof test, the TVB-N is 9.32 mg/100 g which meets the test result, and indicates that after optimization the regression model can be used. The results shows that tea polyphenolson, Nisin, chitosan have an obvious inhibition effect on the spoilage bacteria, isolated from chicken meat. The shelf life of fresh or chilled chicken can be prolonged to 10 days at 4°C.

Acknowledgment: Foundation item: Jiangsu Province Agricultural technology Self-Innovation Fund Project “Innovation and Application of Technology of Slaughter Premium Fresh or Chilled Chicken Chain” (CX (15)1009). National Technology Support Project “Application of Nuclear Technology in Study of Food Security” (2014BAA0305).

References

[1] He Y. The difference of fresh chicken between fresh or chilled chicken [J]. *Encyclopedia*, 2014(6): 56-57.

[2] Liu L, Zhang.D.Q., He.Z.F. The study on the natural preservative to prolong the shelf life of fresh chicken meat product [J]. *S.Food*, 2010, 35(9): 154-159.

[3] Qin.D.F. In HACCP system study on the safety and process of Poultry meat [J].N.Hunan City University, 2015, 24(4): 89-91.

[4] Dong.X.Z, Cai.M.Y. Identification of common bacterial book [M]. Beijing: science press, 2001: 326-365.

[5] Qian.C.R, Huang.Y.X.. Microbiology experiment guide [J].*P.Peking University*, 2000, 207-208.

[6] GB 5009.44-2003. The People's Republic of China Standard, Analysis Method of Standard of Meat and Meat Product Hygiene [J].

[7] Krishnan.K.R, Babuskina.S, Babu P.A.S, et al. Antimicrobial and antioxidant effects of spice extracts on the shelf life extension of raw chicken meat [J]. *J.Food Microbiology*, 2014, 171(8): 32-40.

[8] Divek.V.T. Nair, Kiess.A, Rama..N, Wes.S, et al. The combined efficacy of carvacrol and modified atmosphere packaging on the survival of Salmonella, Campylobacter jejuni and lactic acid bacteria on Turkey breast cutlets [J].*F. Microbiology*, 2015(49): 134-141.

[9] Gao.L, Xie.J, Ye.Z. Response Surface Method Optimization of optimizing the proportion of biological preservatives on fresh or chilled chicken [J]. *F.Industry Technology*, 2015, 22(36): 313-319.

[10] Guo.G.P. The study on preservative technology and spoilage bacteria of roast chicken [D]. YanTai University, 2011.

[11] Li.M.Y. The analysis of bacterial ecology and study on the predicted model of shelf life of fresh meat [D]. NanJing Agriculture University, 2006.

Na Zhao, Xiaoguang Zhou, Huiling Zhou*, Xiangdong Liu

A Method to Evaluate the Energy Consumption, Emission and Cost in China Grain Drying Industry Based on LEAP

Abstract: Grain drying is one of the most energy-consuming and air polluting emission processes in Chinas grain industry, various types of grain drying equipment have been adopted with each of them behaving differently in drying efficiency and pollution emission. Currently, there is no widely accepted method to assess the energy consumption, emissions and cost for the whole grain drying industry, which has brought great difficulties to energy-saving and emission reduction works. In order to optimize the allocation of energy resources, reduce the energy consumption, control the pollution emissions and achieve the purpose of energy saving and emission reduction, this paper introduces the Long-range Energy Alternatives Planning System (LEAP) method for evaluating the energy consumption and emissions of grain drying industry in China. This method combined with the scenario analysis to establish the evaluation model, which can evaluate the energy consumption, emission and cost of the industry and have an influence on the future development and even the national policy adjustment. In this paper, an example has been carried out using the calculation and evaluation process based on LEAP method. The evaluation results show that LEAP-based method can be effectively used to evaluate the energy consumption, emission and economic cost of grain drying industry in China.

Keywords: LEAP Method, energy efficiency assessment, air pollution emission assessment, Scenario analysis, China grain industry.

1 Introduction

The grain drying industry is one of the most important and basic departments in our country, which plays an important role in supporting national grain security. As an important energy consuming sector in agricultural machinery, it's under tremendous pressure to save energy and reduce emissions in the development of new agricultural machinery.

At present, there is a wide range of food drying machines on the market [1-2], and the technology level of the drying process and energy consumption and emissions

***Corresponding author: Huiling Zhou,** Beijing University of Posts and Telecommunications, Beijing, China, E-mail: huiling@bupt.edu.cn
Na Zhao, Xiaoguang Zhou, Beijing University of Posts and Telecommunications, Beijing, China
Xiangdong Liu, China Agricultural University, Beijing, China

are different [3-8]. Experts have put forward theories to assess energy consumption of a single grain drying system: Energy Analysis and Exergy Analysis method [9-10]. However, there isn't any available method to assess the energy consumption, emission and economic cost of the whole grain drying industry.

In other industries, the assessment methods of energy consumption and emission mainly includes: Life Cycle Assessment (LCCA) method which has been applied to cement and public transportation industry [11-13]; Input and Output method which has been applied to tourism industry [14]; CGE model which has been applied to the tourism industry of the various regions in China [15-16], LEAP model which has been applied to logistics, automobile and other industries [17-21].

Based on the studies mentioned above, this paper introduces the Long-range Energy Alternatives Planning System (LEAP) method for evaluating the energy consumption, emission and cost of grain China drying industry.

2 Summary of the Model Based on Leap

LEAP can assess energy consumption, emissions and economic costs through the use of a mathematical model. In addition, combining with the scenario analysis method, LEAP can forecast the industry's energy consumption, pollution emissions and economic costs under the different policies in the future. As shown in Fig. 1, LEAP model consists of four major modules: scenario analysis, driving factors, energy requirements and resources.

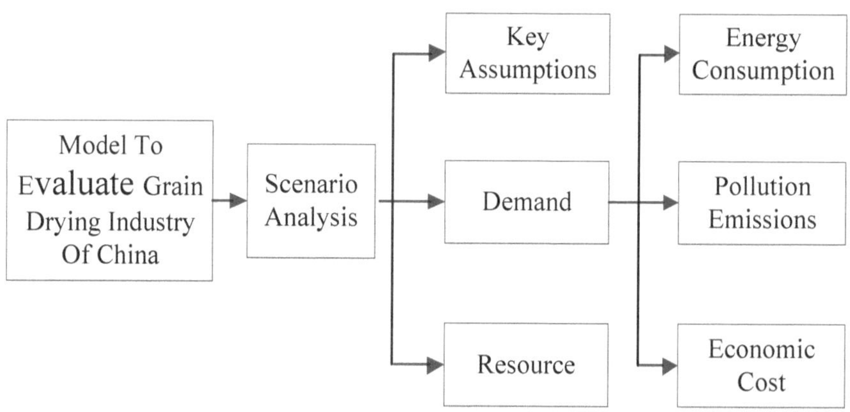

Figure 1: Schematics of China Grain Drying Industry Model based on LEAP

Key Assumptions are factors which relate to energy consumption, emissions and cost of China grain drying industry. They can be divided into basic scenario factors and development scenario factors. The basic scenario factors are according to the status of the grain dryer industry in 2013. The development scenario factor is based on historical development and policy-oriented. The factors in this model include the purchase subsidy of grain dryer, the intensity of national implementation, the proportion of harvesting machinery, grain output, the growth rate of income and GDP, etc.

Demand module is directly related to energy consumption, emissions and cost of the module. This module needs to establish a reasonable calculation of energy consumption, emissions and cost based on the current status of China grain dryer industry and the linkages among various factors and energy consumption, emission and cost, and then establish reasonable data structure in LEAP according to the calculation formula, and finally calculate and analysis to energy consumption, emissions and economic costs.

Resource module is generatated automatically from the above modules, which can directly give the primary and secondary energy consumption of China grain drying industry.

When the data structure of each module has been established, setting the current as the basic scenario. According to the current status of China grain dryer, input of the required data, after which the current grain drying industry energy consumption, emissions and economic costs can be calculated. Same with the basic scenario, the energy consumption, emissions and economic costs of the developing scenarios can be calculated.

3 Leap Model Building

3.1 Theoretical Calculation

According to the investigation and analysis about the current station, the calculation method of energy consumption in grain drying industry is as follows.

$$Device\text{-}Demand(t) = \sum_{h}\sum_{g}\sum_{w}\sum_{e} Act\text{-}Lev(t) * En\text{-}Intensity(t)$$

$$\begin{aligned} Act\text{-}Lev(t) &= AL\text{-}sector(t) * AL\text{-}subsector(t) * \\ &\quad AL\text{-}end\text{-}use(t) * AL\text{-}device(t) \end{aligned}$$

Device-Demand(t):the terminal energy consumption in t years;
Act-Lev(t):the water drying level of mechanical drying;
En-Intensity(t):the energy consumption of drying per water in t years;
AL-sector(t):the activity level of dryer department in t years;
AL-subsector(t):the activity level of grain drying department in t years;
AL-end-use(t):the activity level of terminal water drying in t years;
AL-device(t):the energy consumption level of the equipment in t years;

3.2 Data Structure of Demand Module

To predict the demand, firstly a reasonable data structure should be built. The data structure of the demand can be built up with four gradations. They are sector, sub-sector, terminal equipment and fuel.

When the data structure builds up, the model will calculate the demand in a predictive period with the formula above. The China grain drying model-energy-environment-economy can be built up based on an annual period and the whole nation. The grain parameters were analyzed from historical data. Build up the regression relation between the amount of grain and the driving factors, according to the formula above. Then calculate the energy consumption, pollution emissions and cost in every scenario.

Through the analysis of the current situation of Chinas grain drying industry, the construction of a hierarchical model is as follows: the quantity of dryers, the quantity of drying grain, the water drying level, the per energy consumption, the emissions and economic costs of different energy sources.

1) The lamination of the dryer species: First of all, the grain dryer should be classified according to the amount of dry grain with mechanical drying. As shown in Fig. 2, to establish the lamination model, the grain dryers should be divided into 4 sub-lamination: concurrent flow dryer, concurrent and counter flow dryer, mixed flow dryer, circular drier.

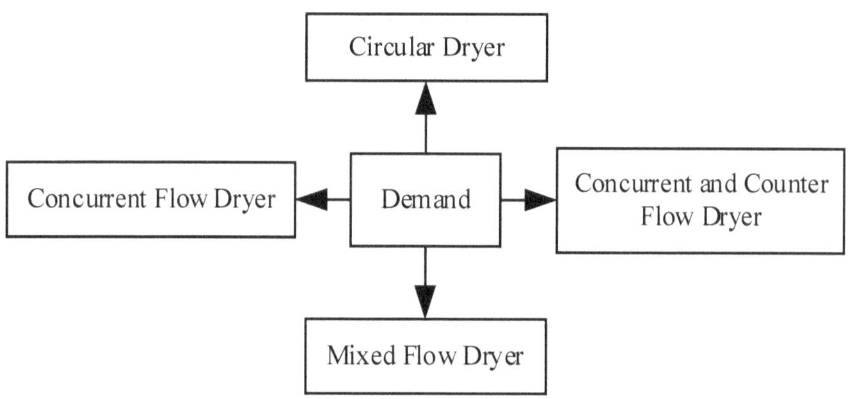

Figure 2: Hierarchical models of different forms of dryer

As drying machine and drying processing are different, each type of dryer section is divided into large, medium and small dryer. Take the concurrent and counter flow dryer as an example, the lamination is shown in Fig. 3.

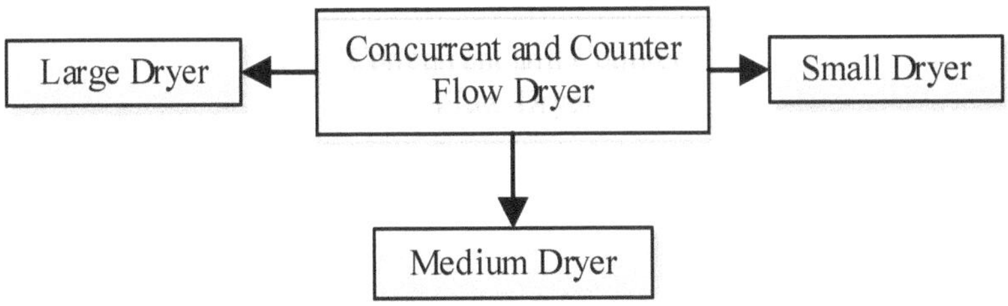

Figure 3: Hierarchical models of different sizes of dryer

Then establish the lamination of different grain dryers - the corn dryer, the rice dryer and the wheat dryer under the lamination of different scale of dryer. Take the middle scale as an example, as shown in Fig. 4.

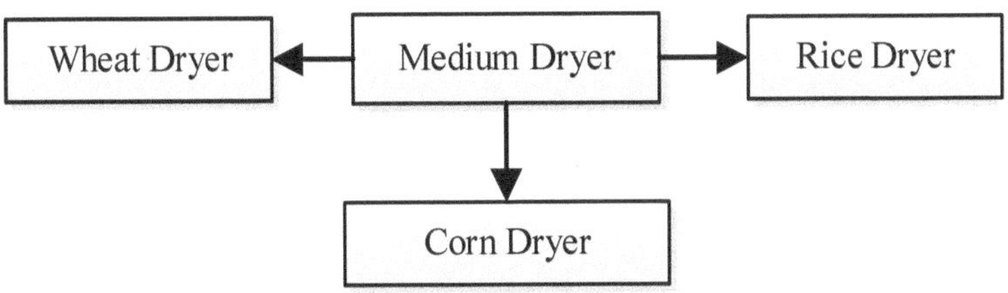

Figure 4: Hierarchical models of medium dryer of different kinds of grain

2) The lamination of the drying quantity: After establishing the lamination of the dryers, the connection between dryers and the drying quantity should be built. We can establish this lamination under the lamination of the corresponding dryer. Take the corn dryer as an example, the structure is shown in Fig. 5.

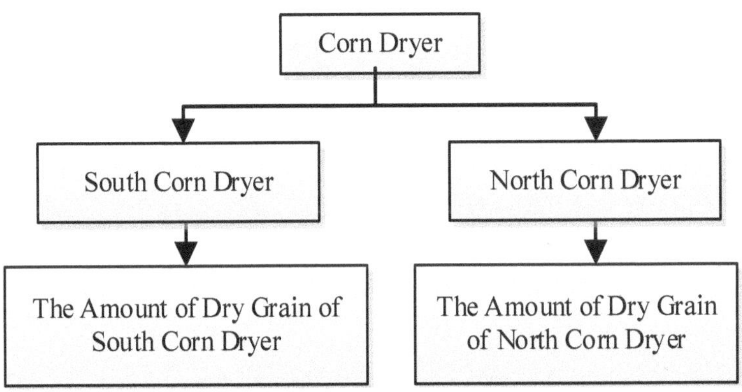

Figure 5: Hierarchical models of different amounts of dry grain of dryer

3) The lamination of different moistures: During the drying process, per water of different moisture of grain, the energy consumption is different. In order to give an accurate assessment, the relation between the moisture and the quantity of grain should be established. Then the evaporation of water can be calculated according to the change of the moisture in the drying process. Take the corn drying as an example, the structure is shown in Fig. 6.

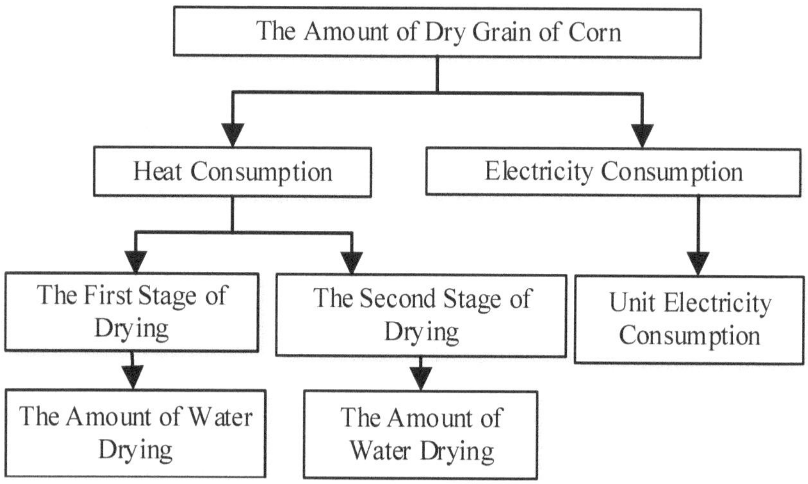

Figure 6: The relation between the water drying and the amount of dry corn

4) The lamination of fuels: During the drying process, we use different fuels, and different fuels have different drying effects, such as energy consumption, emission, price and so on. The relation between the fuels and the moisture is shown as Fig. 7.

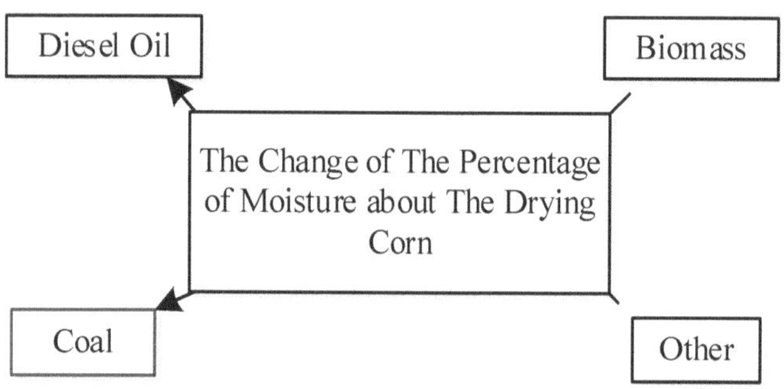

Figure 7: The relation between the change of the percentage of moisture about the drying corn and energy consumption

Different energy sources have different emissions and economic costs. So it is necessary to establish the respective emission factor layers under the respective fuels. As shown in Fig. 8, according to the unit energy consumption, emissions and economic costs of per fuels, the assessment can be achieved.

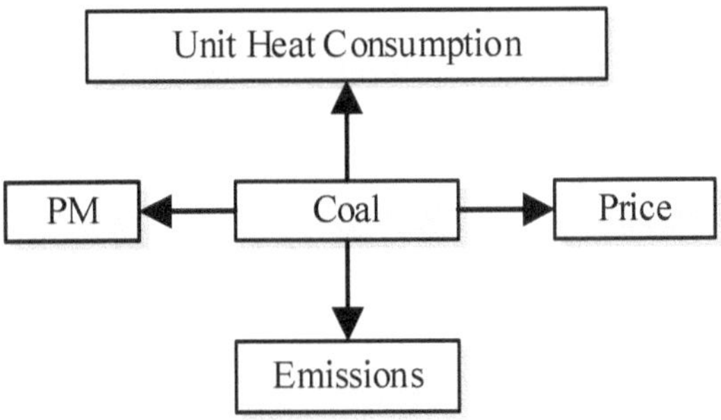

Figure 8: The emissions and economic costs of unit coal

3.3 Scenarios Establishment

The China grain drying model scenarios were classified as the basic scenario and the developing scenario, and were analyzed by a lot of documents about the factors of grain drying equipment, just like the development trend of the national dryer market [22-23], analysis of energy consumption of different kinds of grain dryer [24-26], formative mechanisms of quality [27-29], and the quantity of the grain drying machines and grain [30].

First is the quantitative analysis. According to the current grain drying industry, we set up every parameter of quantitative analysis and calculated the energy consumption, pollution emissions of the basic scenario.

Second is the qualitative analysis. The developing scenarios were set up by different kinds of energy and format of equipment, which focus on a certain type of grain drying system. And then achieve the energy consumption and pollution emissions. Finally, we can propose advice for the development of China grain drying industry in the future from the results of scenario analysis

1. *Basic scenario:* Ignoring policies by the government after 2013, the basic scenario can be set on the base of the currency trend of China grain drying Industry in 2013.
2. *Developing scenario:*

Table 1: Developing Scenario Set

Scenario	Sub-scenario: Fuel type	Content (2013-2020)
1 disseminate concurrent dryer	coal	Increasing the coal drying efficiency and reduce other grain dryers
	Diesel	Increasing the diesel drying efficiency and reduce other grain dryers
	biomass	Increasing the biomass drying efficiency and reduce other grain dryers
2 disseminate concurrent and counter current dryer	coal	Increasing the coal drying efficiency and reduce other grain dryers
	Diesel	Increasing the diesel drying efficiency and reduce other grain dryers
	biomass	Increasing the biomass drying efficiency and reduce other grain dryers
3 disseminate mixed flow dryer	coal	Increasing the coal drying efficiency and reduce other grain dryers
	Diesel	Increasing the diesel drying efficiency and reduce other grain dryers
	biomass	Increasing the biomass drying efficiency and reduce other grain dryers
4 disseminate Circulating Drier	coal	Increasing the coal drying efficiency and reduce other grain dryers
	Diesel	Increasing the diesel drying efficiency and reduce other grain dryers
	biomass	Increasing the biomass drying efficiency and reduce other grain dryers

4 Analysis Example

4.1 Design and Data of the Example

1) Design of the example: The objective of this study is to calculate the energy consumption, emissions and economic cost of China grain drying industry with the LEAP method, so that to verify the feasibility of this method.

The assumption of the study: in order to reduce the data branch, simplify drying mechanical as continuous and circulating [31]. And the basic scenario is 2013 years. And the development scenario is analyzed from historical changes and the policy is same during 2013 to 2020.

2) Data of the example: The data structure in LEAP is based on the above model. The data in LEAP are mainly referenced and derived from the yearbooks. Some of the data obtained from each yearbook are shown in Table 2 and Table 3. Table 2 shows the quantity of different areas. Table 3 gives the quantity of different dryers.

Table 2: The Quantity of Different Areas/ Million Tons

Areas	2013	2012	2011	2010	2009	2008
China	7529	7216	5755	2721	2354	1930
North	5989	5857	4816	1973	1591	1283
South	1541	1359	939	748	763	647

Data from: 2009-2014 "China Agricultural Machinery Industry Yearbook"

Table 3: The Quantity of Different Dryers/ Million Units

	2013	2012	2011	2010	2009	2008
Grain Dryers	4.28	3.60	4.21	3.76	4.33	2.54
Large agricultural machinery	146	132	112	95	78	61
Middle agricultural machinery	381	353	328	297	273	237
Small agricultural machinery	1752	1797	1811	1786	1751	1722

Data from: 2009-2014 "China Agricultural Machinery Industry Yearbook"

Due to the limitations of statistical data, some of the data cannot be referenced directly, therefore it needed to be calculated based on the relevant data. For example, the quantity of continuous and circulating dryer can be calculated from the quantity of north and south in Table 2. The different scale dryers can be calculated from each size of the agricultural machinery in Table 3. The unit heat energy consumption and power consumption can be obtained from the experiment, correcting with the conversion rule between unit heat consumption and processing capacity [26].

Other data sources: the original value of the grain dryer in the model, the total number of grain dryers, the proportion of various types of grain dryers, the quantity of various types grain are referenced from "China Statistical Yearbook", "Chinese Agricultural Statistics" and "China Agricultural Machinery Industry Yearbook". Greenhouse gases and other pollutants emissions are derived from the TED database in LEAP software. The variety of energy prices are set according with the state regulations on industrial energy price.

4.2 Design and Data of the Example

After the model is established, the calculation results of energy consumption, emission and cost in LEAP are as following.

1) The Energy Consumption: The energy consumption in the current scenario is shown in Fig. 9. In 2013, the total energy consumption is 1.7795 million tons of standard coal, and the consumption of coal is the highest, reaching 1.4774 million tons, 83% of the total energy consumption, other energy accounts for 17%.

The current literature only gives the results of energy consumption, and cannot be traced back to the original calculation process. The calculation method in this paper takes a number of factors into account, and the data structure will be more precise. Therefore, the evaluation method based on LEAP may be more accurate. However, this is an example to show the calculation process of energy consumption, emission and cost based on LEAP model, and the current research data is not comprehensive enough, so the results of this study may be a little deviation.

In addition, maintaining the current drying method unchanged, the total energy consumption is expected to a linear increase. The total energy consumption will increase to 2,503,900 tons, an increase of 40.7% by 2020. The average annual growth is 5.8%.

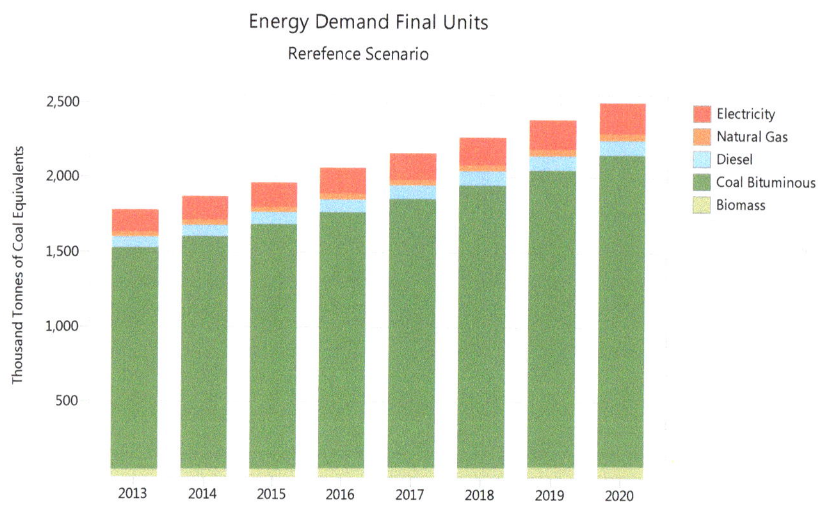

Fuels	2013	2014	2015	2016	2017	2018	2019	2020
Electricity	143.2	150.3	157.9	165.7	174.0	182.7	191.9	201.5
Natural Gas	32.7	34.3	36.0	37.8	39.7	41.7	43.8	46.0
Diesel	73.8	77.5	81.4	85.4	89.7	94.2	98.9	103.8
Coal Bituminous	1,477.4	1,551.3	1,628.9	1,710.3	1,795.8	1,885.6	1,979.9	2,078.9
Biomass	52.4	55.0	57.8	60.7	63.7	66.9	70.2	73.8
Total	1,779.5	1,868.4	1,961.9	2,060.0	2,163.0	2,271.1	2,384.7	2,503.9

Figure 9: Grain drying industry energy consumption

2) The pollutant emissions: As shown in Fig. 10, select Pollutant Emissions in the current scenario. In 2013, the total pollutant emission of grain drying reached 4.56 million tons of coal equivalent, of which the largest amount is CO_2, reaching 422.33 million tons, accounting for 92.6%. To maintain the current development situation, the pollutant emissions must sustain a linear growth. By 2020, the total emissions will reach 6.41 million tons of standard coal equivalent, an increase of 40.7%, consistent with the growth in energy consumption.

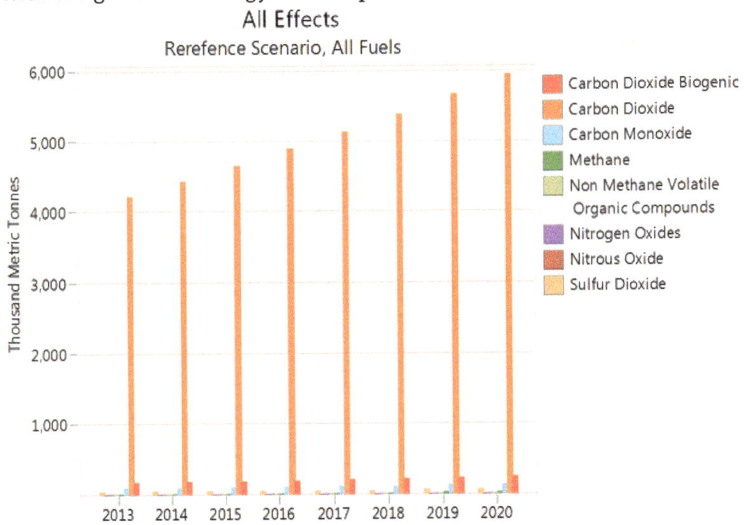

Effects	2013	2014	2015	2016	2017	2018	2019	2020
Carbon Dioxide Biogenic	168.3	176.7	185.6	194.8	204.6	214.8	225.5	236.8
Carbon Dioxide	4,223.3	4,434.5	4,656.2	4,889.0	5,133.5	5,390.2	5,659.7	5,942.7
Carbon Monoxide	95.6	100.4	105.4	110.7	116.2	122.0	128.1	134.6
Methane	13.5	14.2	14.9	15.6	16.4	17.2	18.1	19.0
Non Methane Volatile Organic Compounds	10.1	10.6	11.1	11.7	12.3	12.9	13.5	14.2
Nitrogen Oxides	7.8	8.2	8.6	9.0	9.5	9.9	10.4	10.9
Nitrous Oxide	0.1	0.1	0.1	0.1	0.1	0.1	0.1	0.1
Sulfur Dioxide	41.3	43.4	45.6	47.8	50.2	52.7	55.4	58.1
Total	4,560.0	4,788.0	5,027.4	5,278.8	5,542.7	5,819.8	6,110.8	6,416.4

Figure 10: Grain drying industry emissions

3) The economic cost: Select Social Cost in Costs in the current scenario, as shown in Fig. 11. The total social cost of grain drying equipment in 2013 reached 406 million yuan, of which the social cost of the continuous dryer was 234 million yuan, accounting for 57.6%. However, the larger continuous dryer dried the most of the grain, but only accounted for half the cost. To maintain the current development situation, by 2020, the total cost will be 571 million yuan, an increase of 143.6%, an average annual growth of 20.5%.

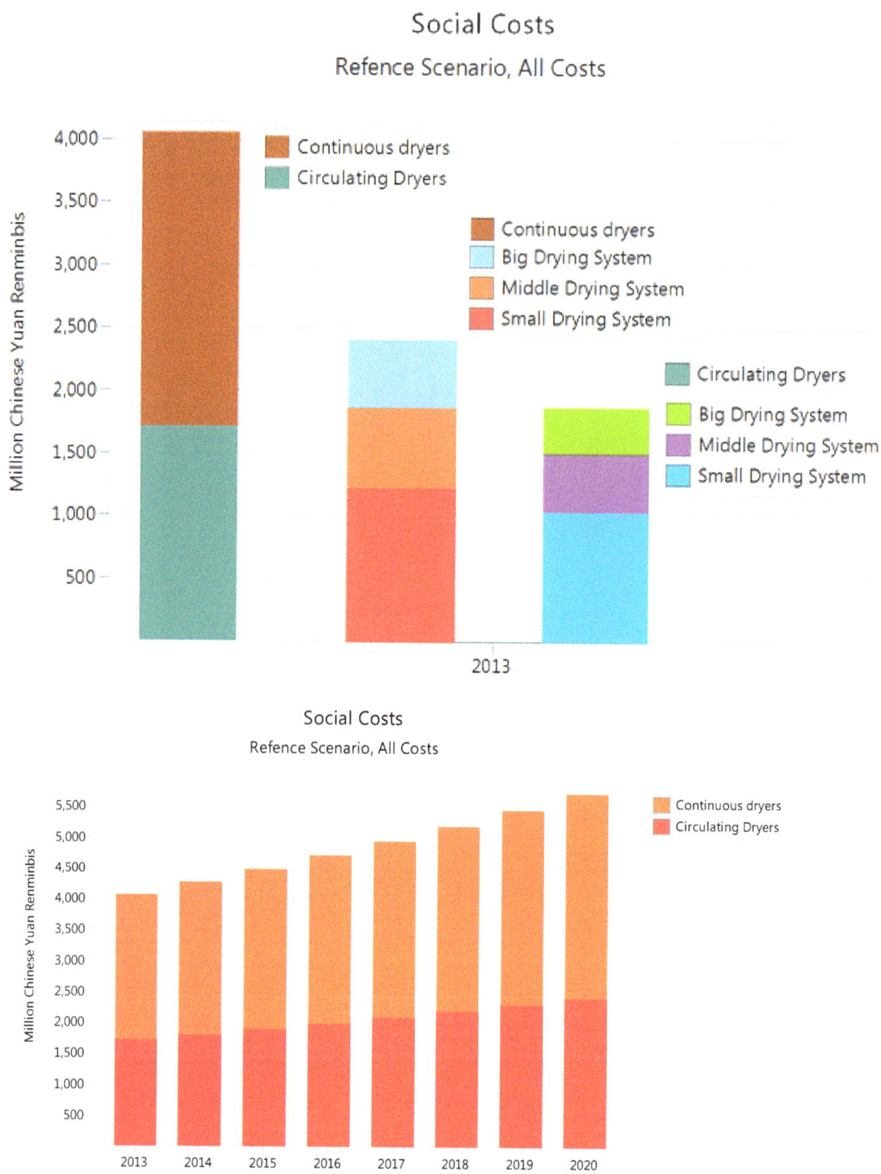

Branches	2013	2014	2015	2016	2017	2018	2019	2020
Continuous dryers	2,339.9	2,456.9	2,579.8	2,708.8	2,844.2	2,986.4	3,135.7	3,292.5
Circulating Dryers	1,721.4	1,807.5	1,897.9	1,992.8	2,092.4	2,197.0	2,306.9	2,422.2
Total	4,061.4	4,264.4	4,477.6	4,701.5	4,936.6	5,183.4	5,442.6	5,714.7

Figure 11: The cost of grain drying industry

5 Conclusions and Prospect

It can be concluded that LEAP is very suitable for energy consumption, emission and economic cost assessment of grain drying through the analysis of the process based on the LEAP calculation method and example analysis. In addition, we can conclude as follows:

(1) At the end of 2013, the energy consumption amounted to 1,795,900 tons standard coal, the majority is coal; the emissions amounted to 4 million 560 thousand tons, the CO_2 occupied the largest proportion; the cost amounted to 406 million yuan. The Large agricultural machinery is the lowest costs.

(2) Keeping the current development situation, the energy consumption, emissions and costs of the drying industry will be linear growth. By 2020, the energy consumption and emissions rise 40.7%, the cost rose higher, rise to 143.6%.

The data is not very perfect in this paper. In the future research, more date will be evaluated to study energy consumption, emissions and cost of grain drying.

In addition, the LEAP model cannot only calculate the current situation, but also the changed development scenarios, which can predict the status of energy consumption, emissions and cost. Therefore it can analyze the influence about some policies on the drying industry

Acknowledgment: 1. Special Fund for Scientific Research of Public Welfare Industry of State Grain Bureau (No. 201413006); 2. Research on Special Purpose Fund for Grain Public Welfare Industry (No.201513002).

References

[1] Yang Y., Mou G.L., Zhang X.J., et al. Current situation and development trend of domestic grain dry[J]. Xinjiang Agricultural Mechanization, 2013, 04: 23-25.
[2] Jin Z.. Types of rice drying equipment[J]. Hunan Agricultural Machinery, 2014, 08: 157.
[3] Xie Q.Z., Liu J, Shi J.F., et al. Experimental tudy on continuous mixed-flow drying technology for rice[J].Transactions of the Chinese Society of Agricultural Engineering(Transactions of the CSAE), 2006, 22(3): 129-132.(in Chinese with English abstract)
[4] Cao C.W., Zhu W.X.. Computer simulation of drying process of agricultural products[M]. Beijing: China Agriculture Press, 2001.
[5] HuW.L., Li C.Y., Xu F.Y.. Experimental study on fast dry craft of rough rice Thin Layer[J]. Transactions of the Chinese Society for Agricultural Machinery, 2007, 38(4): 103-106 (in Chinese with English abstract)
[6] Zheng X.Z., Xia J.Q.,Yang Y.Q.. Experimental study on paddy drying technology by using humidified and heated air[J].Transactions of the Chinese Society of Agricultural Engineering (Transactions of the CSAE), 2000, 16(3): 81-83. (in Chinese with English abstract)
[7] Wang G.X. Study on process of rice concurrentmixed flow drying[J]. Transactions of the Chinese Society of Agricultural Engineering (Transactions of the CSAE), 2000, 16(2): 109-112. (in Chinese with English abstract)

[8] Shen B.Q.. Research on grain drier in Northeast China[J]. Grain Distribution Technology, 2007, 01: 19-21.
[9] Li C.Y.. Energy evaluation theory of hot air drying system for grains[J]. Transactions of the Chinese Society of Agricultural Engineering (Transactions of the CSAE), 2012, 28(12): 1-6. (in Chinese with English abstract)
[10] Li C.Q. Exergy evaluation theory of hot air drying system for grains[J]. Transactions of the Chinese Society of Agricultural Engineering, 2012, 12: 1-6.
[11] Lin X.D., Bin L., Tian L.. Analysis of energy conservation and greenhouse gas emissions reduction potential of buses in Beijing city based on life cycle assessment[J]. Acta Scientiae Circumstantiae, 2015, 02: 576-584.
[12] Zhang X., Jiang K.J.. Research on the analysis of all round consumption of cement products in Beijing cement plant[J]. Macroeconomics, 2008, 02: 18-22+28.
[13] Wang X.Q., Liang D.L., Wang X.D.. Estimation of greenhouse gas emissions from dairy farming systems based on LCA[J]. Transactions of the Chinese Society of Agricultural Engineering, 2012, 13: 179-184.
[14] Zhong Y.D., Shi S.Y., Li S.H., et al. Empirical research on measurement framework construction for tourist industry carbon emission in China: A perspective of input-output[J]. Journal of Central South University of Forestry & Technology, 2015, 01: 132-139+144.
[15] Song T., Niu Y.F.. Study on the Oversea Evaluation of Economic Impact of Tourism Based on Computable General Equilibrium(CGE)Model[J]. Tourism Tribune, 2008, 10: 23-28.
[16] Du Y.M.. Study of CGE Model for Fujian Province[D]. Fuzhou University, 2004.
[17] Lu Y.. Study on the potential of carbon emissions from the logistics industry based on LEAP[D]. Beijing Jiaotong University, 2011.
[18] TiS.G.. Energy and environment scenarionalysis for resident sector based on LEAP[D]. Henan Agriculture of University, 2010.
[19] ShiY.C.. Analysis of Dalian Industrial Energy Consumption Based on LEAP Model[D]. Dalian Maritime University, 2014.
[20] Pan P.F., Wang P., LI W., et al. Analysis of the energy-saving and emission-reduction potential for transportation in Henan Province based on LEAP[D]. Henan Agricultural University, 2014.
[21] Wang K., Wang C., Lei X.D., et al. Abatement potential of CO_2 emissions from China's iron and steel industry based on LEAP[J]. Journal of Tsinghua University(Science and Technology), 2006, 12: 1982-1986.
[22] Chen J.G., Cao Y.G., Zhang L.L., et al. Analysis and Countermeasures on the Ecosystem Questions of Agricultural Machinery[J]. Journal of Agricultural Mechanization Research, 2007, 09: 9-11+15.
[23] Peng Y.K.. The development status and trend of grain drying technology in China[J].Farm Products Processing, 2013, 05: 6-7.
[24] Xing B.L., Liu Q.G.. Evaluation of Large Grain Dryers in China[J]. Transactions of The Chinese Society of Agricultural Engineering, 1991, 02: 72-79.
[25] Pan J.J.,Yin X.H.. Study on the Conversion Rules of Heat Consumption and Processing Units of Grain Dryer[J]. Drying Technology & Equipment, 2008, 02: 65-70+101.
[26] GB/T 21162-2007. Conversion rule for specific heat consumption and throughput of concurrent—Flow grain dryer[S].
[27] Naret Meeso, Adisak Nathakaranakule, Thanid Madhiyanon, et al. Modelling of far-infrared irradiation in paddy drying process[J]. Journal of food engineering, 2007, 78(4): 1248-1258.
[28] Ragab Khir, Pan Z.L., Adel salim, et al. Moisture diffusivity of rough rice under infrared radiation drying[J]. LWT-Food Science and Technology, 2011, 44(1): 1126-1132.
[29] Zheng X.Z., Zhao X.D., Chen L.. Effect of drying conditions of paddy on rice taste and quality[J]. Transactions of the Chinese Society of Agricultural Engineering (Transactions of the CSAE), 2000, 16(4): 126-128. (in Chinese with English abstract)

[30] The Editorial Projects in China Agricultural Machinery Industry Yearbook. China Agricultural Machinery Industry Yearbook[M]. Beijing: China Machine Press, 2009-2013.
[31] Hao Y.. The applications and development countermeasures of grain dryer[J]. Advanced Agricultural Equipment, 2014, 04: 55-58

Haiyan Wang*, Lei Zhang, Shuiyuan Cheng, Fan Feng, Wei Liu

Isolation of Candida Pseudolambica and Characterization of Its Hydrogen Production

Abstract: In order to find more effective biohydrogen producing microorganisms, a fermentative hydrogen-producing strain L3 was isolated and identified as *Candida pseudolambica* (*C. pseudolambica*). The optical nitrogen source and metal for the growth and increased hydrogen production of the yeast were investigated in an anaerobic reactor using glucose and starch as substrates. The maximum hydrogen yield of 3.19 mol H_2/mol glucose was obtained fed with ammonium sulfate and Fe^{3+} ions in the medium. It was also concluded that for *C. pseudolambica* L3 strain that hydrogen production was several times higher under a dark environment than that under diurnal light/night conditions. It appears to be a new and better choice for highly efficient dark fermentative hydrogen production using *C. pseudolambica* L3.

Keywords: biohydrogen production, *Candida pseudolambica*, isolation, characterization, metabolic production.

1 Introduction

The energy and global environmental crises have encouraged scientists to develop non-polluting and renewable energy sources [1-4]. Compared with the use of fossil fuels, bio-hydrogen production is gaining attention based on its clean burning, versatile, and renewable characteristics [5-7]. The process of biological hydrogen production is carried out at lower temperatures and pressures, and is therefore less energy-intensive than thermochemical and electrochemical processes. A number of different light-dependent and fermentative biological processes may be employed in the production of hydrogen including direct biophotolysis such as green microalgae [8], indirect biophotolysis such as cyanobacteria [9], photo-fermentation such as purple non-sulfur bacteria [10] and dark fermentation such as *Bacillus* [11] and *Clostridium* [12]. In the direct biophotolysis process, a direct transfer of electrons from water to hydrogen-evolving enzymes is accompanied by oxygen and hydrogen evolution. Indirect biophotolysis process involves separation between the water

***Corresponding author: Haiyan Wang,** Key Laboratory of Beijing on Regional Air Pollution Control, Beijing University of Technology, Beijing, China, E-mail: why@bjut.edu.cn
Lei Zhang, Shuiyuan Cheng, Fan Feng, Key Laboratory of Beijing on Regional Air Pollution Control, Beijing University of Technology, Beijing, China
Wei Liu, College of Life Science and Bioengineering, Beijing University of Technology, Beijing, China

splitting reaction (aerobic stage) and hydrogen production (anaerobic stage). In dark fermentation, carbohydrates are converted to H_2, CO_2 and organic acids.

Some scientists have studied the production of bio-hydrogen using pure culture from carbohydrates, including species of *Enterobacter* [13], *Bacillus* [11] and *Clostridium* [12]. Oh et al. [14] studied anaerobic glucose metabolism and hydrogen production with a newly isolated chemoheterotrophic bacterium, *Citrobacteramalonaticus*Y19 and found that its maximum H_2 production was 8.7 mol H_2/mol glucose. Kumar and Das [15,16] isolated *Enterobacter cloacae* IIT-BT08 and *Enterobacter aerogenes* DM11 to yield hydrogen, and the maximum hydrogen production capacity reached 2.7 and 2.80 mol H_2/mol hexose, respectively, in static culture conditions. Xing et al. [17] isolated efficient ethanol-type hydrogen-producing bacteria *Ethanoligenensharbinense* sp. R3 and *Ethanoligenens harbinense* sp. Y3 consecutively from a continuous stirred tank reactor. When using glucose as the substrate, the maximum hydrogen production was 35.74 mmol/g/h with dry cells and 2.81 mol H_2/mol glucose. However, the research group [18] isolated *T. thermosaccharolyticum* W16 from hot springs, which can effectively produce hydrogen using xylose as the fermentation substrate and found that the maximum hydrogen production rate was 10.7 mmol/L/h and 2.19 mol H_2/mol xylose.

In this study, a new hydrogen-producing microorganism was isolated. The morphological and biochemical tests, as well as DNA identification and the characterization of hydrogen yield are presented.

2 Materials and Methods

2.1 Isolation of Inoculums

The sludge was obtained at the Liuminying biogas fermentation pond in Beijing, China. The pure microorganism was separated and cultivated using identical media with the following composition:

- casein peptone, 10 g/L;
- glucose, 5 g/L;
- compound sodium chloride, 5 g/L;
- sodium acetate, 3 g/L;
- agar powder, 0.5 g/L;
- yeast extract powder, 3 g/L;
- beef extract powder, 10 g/L;
- cysteine hydrochloride, 0.5 g/L; and soluble starch, 1 g/L.

2.2 Reactor Experiments and Hydrogen Production

The hydrogen production experiments were performed within 150 mL serum vials as batch reactors. A mixture of the seed microorganism, which was composed of the substrate and 10 mL of a 24 h culture of the isolated microorganism, was filled to 100 mL.

The bottles were incubated at 35°C in an orbital shaker with a rotation speed of 150 rpm to provide better contact among substrates.

2.3 Analytical Methods

The gas composition was analyzed using a gas chromatograph with a thermal conductivity detector (SP-3420, China). The operational temperatures of the injection port, the oven and the detector were 100°C, 80°C and 120°C, respectively.

The volatile fatty acids (VFAs) were also determined using a gas chromatograph that was equipped with a flame ionization detector. The initial temperature of the column was 130°C and then increased at a rate of 10°C/min to reach a final temperature of 240°C. The temperature of both the injector and the detector was 250°C. Depending on the expected biogas production, biogas production was measured in 50 mL plastic syringes, which were fitted with hypodermic needles as described previously by Owen et al. [19,20] at 35°C and 760 mmHg.

Five anions (SO_4^{2-}, NO_3^-, Cl^-, F^-, NO_2^-) and five cations (NH_4^+, Ca^{2+}, K^+, Mg^{2+}, Na^+) were analyzed by Ion Chromatography (IC, model, Dionex1100), which consists of a separation column (DionexIonpac AS11 for anion and CS12A for cation), a guard column (DionexIonpac AG 11 for anion and CG12A for cation), and a electrolytically regenerated suppressor (DionexIonpacERS500). The samples were filtered through microporous membranes (pore size 0.45 μm; diameter 25 mm) to remove the insoluble substances. After passing through the microporous membranes, each filtrate was stored at 4°C in a clean tube for analysis.

2.4 Morphological and Biochemical Test

The morphology of the isolated microorganisms was studied using SEM (S-4300, Hitachi) and light microscopy.

2.5 Genomic DNA Extraction

A small amount of vigorously growing microorganism cells were obtained from the inclined plane test and suspended in 300 μL lysate (100 mmol/L Tris, 30 mmol/L

EDTA, 0.5% SDS, pH8).This sample was then incubated in a water bath at 100°C for 15 minutes. After cooling, 300 μL of 2.5M potassium acetate (pH 7.5) was added, and the sample was put on ice for 1 hour. Then, the sample was centrifuged at 13000 r/min for 5 minutes. The same volume of chloroform-isoamylalcohol (24:1) was added to the supernatant, vortexed and centrifuged at 13000 r/min again. This step was repeated twice. Subsequently, the same volume of pre-chilled isopropanol was added and the sample was stored at -20°C for 15 minutes. Next, the sample was centrifuged at 13000 r/min for 7 minutes. The precipitate was then washed with 100 μL 70% ethanol and vacuum dried. Finally, 50 μL sterile water was added to re-suspend the pellet, and the sample was stored at -20°C until use.

2.6 26S rDNA Cloning and Sequencing

PCR amplification was performed in 100 μL reaction mixtures that contained 2.5 nanograms of template, which were total DNA, 10×PCR buffer (Mg^{2}-Plus) 10 μL, 5U Taq polymerase 0.8 μL, 2.5 mmol/L dNTPs 8 μL, 10 μmol/L of F1 (NL-1: 5'-GCA TAT CAA TAA GCG GAGGAA AAG-3') and F2 (NL-4: 5-'GGT CCG TGT TTC AAG ACG G-3') primers 2 μL each, and ddH_2O.

The thermal cycling conditions that were used for the amplification of the rDNA gene of the bacteria were 30 cycles of 94°C and 52°C each for 1 min and 72°C for 1 min 30 s.

3 Results and Discussion

3.1 Isolation and Characterization of The Microorganism

The SEM and microscope pictures of the isolate L3 are shown in Fig. 1, and the morphological features are listed in Table 1.

Table 1: Morphological features

	Cell shape	Cell length (μm)	Colony color	Endospore	Crystal	Motility
L3	spherical	20	cream	+	-	+

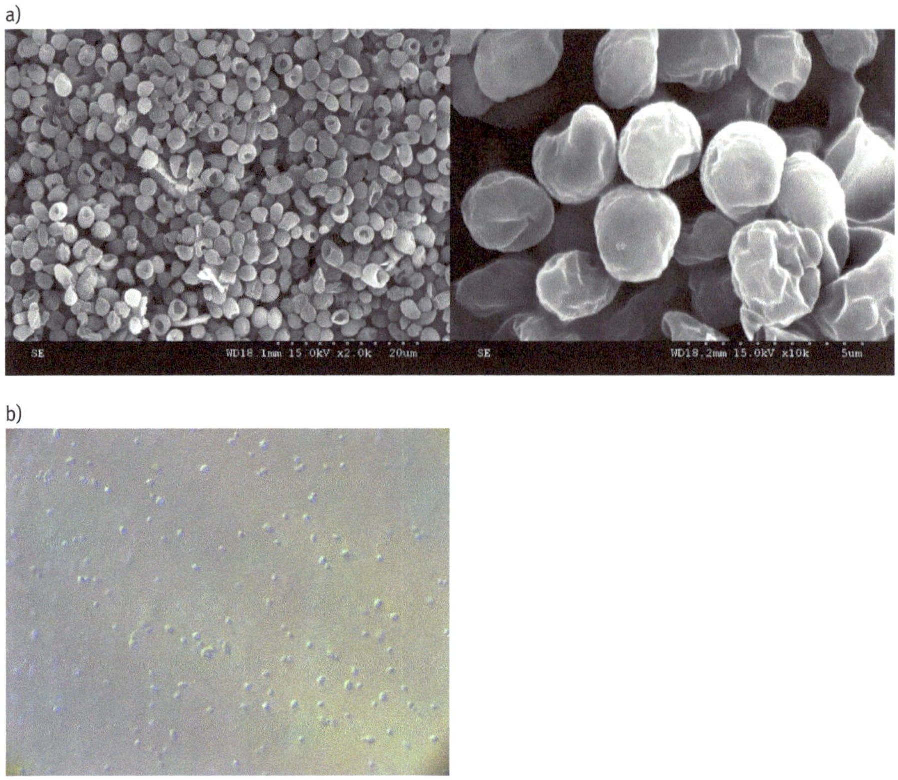

Figure 1: a. SEM picture of the isolate L3 b. microscope picture of the isolate L3

3.2 26SrDNA Extraction and PCR Amplification

The 26S rDNA (D1/D2 region) sequence of L3 is shown in Fig. 2, and the phylogenetic tree is shown in Fig. 3. The BLAST analysis of the 26S rDNA sequence and the morphological features of the microorganism suggested that strain L3 belongs to *Candida pseudolambica*. The 26 SrDNA sequence of strain L3 shares 99.5% and 97.7% identity with the sequences of *C. pseudolambica* NRRL Y-17318T (EF550235) and *C. californica* NRRL Y-27254T (EF550230), respectively. The phylogenetic analysis was performed on the sequence of strain L3 and its closest relatives using the MEGA 4.1 software. Strain L3 was closely related to *Candida pseudolambica*. The biochemical characteristics of strain L3, which are shown in Table 2, also supported its assignment as *C. pseudolambica*.

1 TGCATATCCA AAAAGCGGAG GAAAAGAAAC CAACAGGGATTGCCCCAGTAGCGGCGAGTG
61 AAGCGGCAAG AGCTCAGATT TGGAATCGTG TTTCGGCACG AGTTGTAAAT TGCAGGTTGG
121 AGTCTTTGTG GAAGCGTGTG TCCAAGTCCC TTGGAACAGG GCGCCATTGA GGGTGAGAGC
181 CCCGTGGGAT GCGCGCGGAA GCAACAAGGC CCTTCTGACG AGTCGAGTTG TTTGGGAATG
241 CAGCTCTAAG CGGGTGGTAA ATTCCATCTA AGGCTAAATA CTGGCGAGAG ACCGATAGCG
301 AACAAGTACT GTGAAGGAAA GATGAAAAGCACTTTGAAAAGAGAGTGAAACAGCACGTGA
361 AATTGTTGAA AGGGAAGGGT ATTGGGCCCG ACATGGGGAT TGCGCATCTC TGCTCCTTGT
421 GGGCGGGGCG CTGTGCTTTT CCTGGGCCAG CATCGGTTTT TGCCGCAGGA GAAGGGGTAC
481 TGGAATGTGG CTCCTCGGAG TGTTATAGCC AGTGCCAGAT GCTGCGTGCG GGGACCGAGG
541 GCTGCGACAT CTGTCTCGGA TGCTGGCACA ACGGCGCAAT ACCGCCCGTC TTGAAACACG
601 GGACCAAAA

Figure 2: 26S rDNA (D1/D2 region) sequence of strain L3

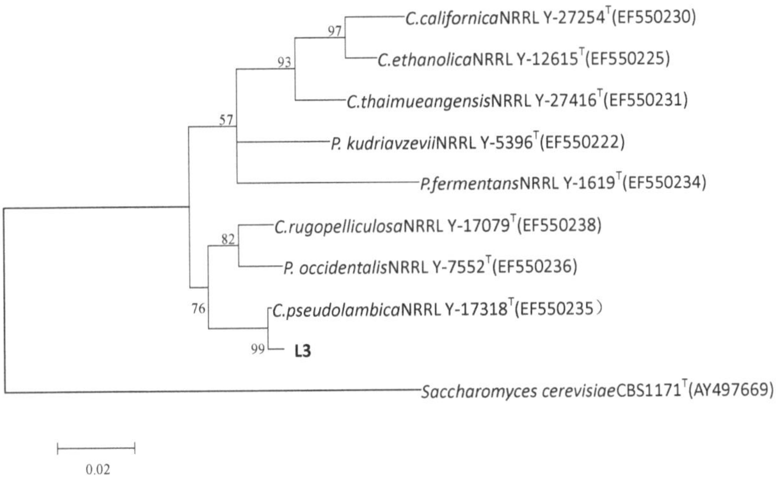

Figure 3: Neighbor-joining tree showing the phylogenetic correlation between the hydrogen-yielding yeast isolates (L3) and representative species (GenBank accession numbers in parentheses). The bar represents 0.02 substitutions per site, and bootstrap values (n = 1000) are displayed at the nodes (*C., Candida; P.,Pichia*)

Table 2: The biochemical test results

Temperatures	Growth tests	Carbon source reaction			
at 25ºC	yes	D-Glucose	+	Ethylamine	+
at 30ºC	yes	D-Galactose	-	L-Lysine	+
at 35ºC	yes	L-Sorbose	-	D-Tryptophan	+
at 37ºC	no	D-Glucosamine	+	Imidazole	-
at 40ºC	no	Starch	-		

3.3 Effect of Nitrogen Source and Metal Ions on Hydrogen Production

Nitrogen is the most commonly reported nutrient-limiting factor in the growth and lipid accumulation. We conducted experiments to investigate the effect of nitrogen sources and dark environment on cumulative hydrogen production and VFA concentrations. Tests were performed with 25 mL tubes which contained 10 mL of lysogeny broth (LB) culture. 3 mmol/L of urea, ammonium oxalate, ammonium sulfate and sodium nitrite were tested as nitrogen sources. Different nitrogen sources resulted in great differences on gas production. Experimental results indicated that a suitable nitrogen source could facilitate cell growth and hydrogen evolution. Fig. 4(a) shows clearly that ammonium sulfate was the optimal choice for hydrogen production in strain L3 as the medium supplemented with it and under dark environment obtained the highest hydrogen yield of 1.52 mol H_2/mol glucose, while the second highest hydrogen yield of 0.37 mol H_2/mol glucose was obtained in the presence of sodium nitrite. When urea and ammonium oxalate were tested, the hydrogen yield were even lower than that in control reactors leading to the conclusion that they were not suitable for the growth of strain L3 and its hydrogen production.

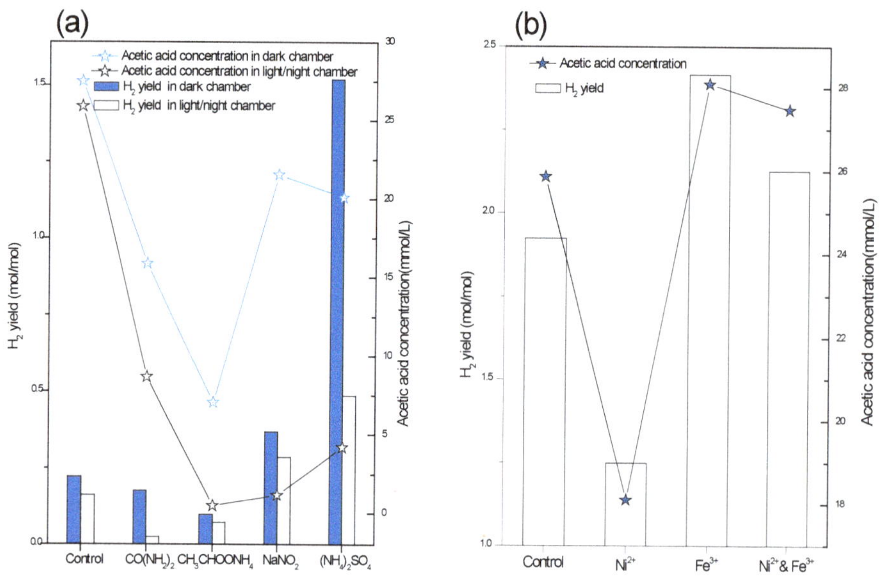

Figure 4: Hydrogen production profiles of strain *C. pseudolambica* L3 grown on medium supplemented with different nitrogen source and metal ions.
(a) Profiles of nitrogen source for hydrogen production and acetic acid concentration.
(b) Profiles of metal ions for hydrogen production and acetic acid concentration.

Another interesting phenomenon was found that hydrogen production was several times higher under a dark environment than that under diurnal light/night conditions with the same medium. According to this result, strain L3 was more effective in dark fermentations than in light-dependent biohydrogen production.

Liquid fermentation of end products was measured. Results showed acetic acid was the main product and less ethanol and propanoic acid were also examined. Fig. 4(a) shows the acetic acid concentration and its varying level under dark vs. light/night conditions. It was obvious that acetic acid concentration was higher under dark than under diurnal conditions compared with the same nitrogen source and the trend was similar with hydrogen production. It could be concluded that more hydrogen production was obtained and more acetic acid was yielded with same medium under dark conditions.

The fermentation of glucose to biohydrogen has a concentration of Ni^{2+} ions and Fe^{3+} ions, both of these ions were supplemented in the medium shown in Fig. 4(b). The results confirmed that strain *C. pseudolambica* L3 favoured with Fe^{3+} ions leads to higher hydrogen yield. The initial Ni^{2+} ions concentration of 30 μmol/L reduced the hydrogen yield by 35%, in comparison with the control test. With regard to 30 μmol/L Fe^{3+} ions and co-metal of Ni^{2+} ions and Fe^{3+} ions test, the hydrogen yield was enhanced by 26% and 11%, respectively, as compared to the control test. The enhancement might be due to the interaction of the Fe^{3+} ions and the [FeFe]-hydrogenase-like genes [21] may be expressed in *C. pseudolambica* L3. The genome of some yeast encodes the essential protein Nar1p that is conserved in virtually all eukaryotes and shows striking sequence similarity to bacterial iron-only hydrogenases. The presentation of co-metal of Ni^{2+} and Fe^{3+} ions led to the positive effect on hydrogen evolution compared with the Ni^{2+} ions test but negative effect compared with the Fe^{3+} ions test as for the proposed expressed Nar1p in *C. pseudolambica* L3 was dedicated to functioning in hydrogen formation. Supplementation of Ni^{2+} ions reduced the biohydrogen yield to 25 mmol/L. Supplementation of Fe^{3+} ions enhanced the biohydrogen yield due to more hydrogen evolution of [FeFe]-hydrogenase-like genes expressed in *C. pseudolambica* L3, resulting in the highest biohydrogen yield of 49 mmol/L medium which was 68% higher than that of the Ni^{2+} ions test. [FeFe]-H_2ases are the most active H_2 producers [22]. They contain a complex cofactor, the "H-cluster", consisting of a simple cubane [4Fe-4S]-cluster bound via cysteine to a unique [2Fe]-subcluster in which two [Fe]-atoms are coordinated by CO, CN^- and a dithiolate. As for the metabolism production it could be seen that higher hydrogen yields were achieved at higher concentration of acetic acid.

3.4 Biohydrogen Production of Strain *C.pseudolambica* L3 under Dark Environment

In order to better understand the hydrogen yield of the strain *C. pseudolambica* L3 under dark conditions, two groups of comparable tests with supplementation of 30 μmol/L Fe^{3+} ions and 3 mmol/L ammonium sulfate were performed. Both group tests were carried out in sealed vials having a total capacity of 150 mL, with 100 mL medium used and under dark and diurnal light/night condition, respectively. Hydrogen and liquid end products were investigated. Fig. 5(a) depicts the time course of variation of hydrogen productivity and accumulative hydrogen evolution by strain *C. pseudolambica* L3 when the initial glucose and starch concentration were 5 g/L and 1 g/L, respectively. Nearly 480 ml and only 190 ml cumulative hydrogen were obtained in dark and light/night chambers respectively at the end of 240 hours. At the beginning of 48th hour, H_2 production in natural light reactor was 2 times that in dark reactor which was 108 mL and 58 mL, respectively. The next day, H_2 production increased sharply and then decreased gradually till the end of the reaction under dark

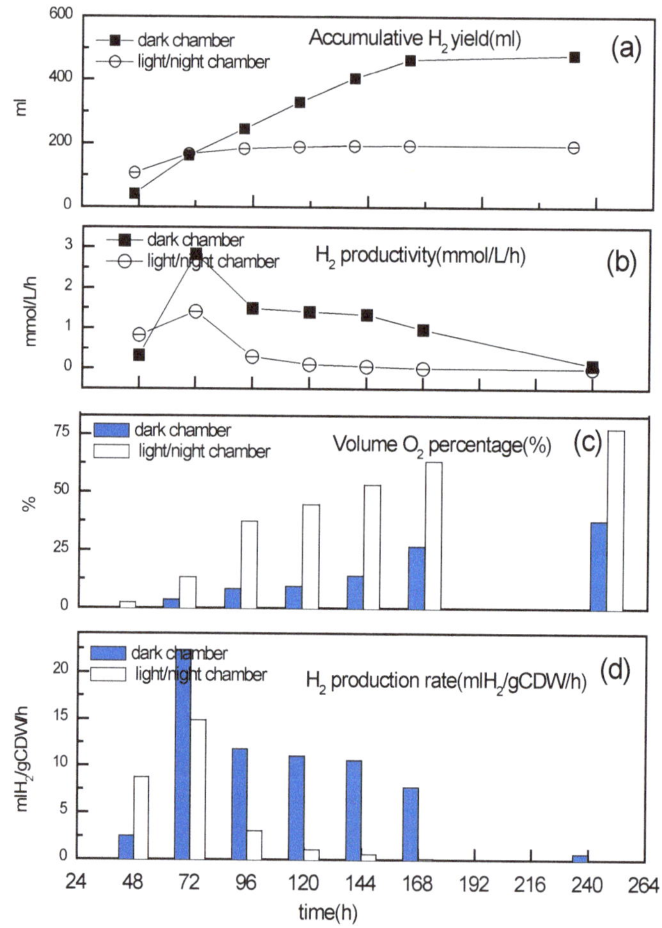

Figure 5: Time course of fermentation profile of strain *C. pseudolambica* L3 grown on glucose and starch batch cultivation (a) Time course of accumulative H_2 yield; (b) Time course of H_2 productivity; (c) Time course of volume O_2 Percentage; (d) Time course of H_2 production rate.

conditions, while H_2 yield began to decrease accordingly. As a result, the hydrogen yield attained to 3.19 mol H_2/mol glucose in the dark reactor which was 4 times the hydrogen yield in the light/night reactor, although H_2 production rate in the light/night reactor was higher at the beginning of the reaction. Hydrogen production type of *C. seudolambica* L3 strain may be attributed to indirect biophotolysis and the existence of [FeFe]-hydrogenase-like genes. In indirect biophotolysis, reducing equivalents for hydrogen production can be generated either in light or in dark conditions.

Fig. 5(b) describes the relation between oxygen volume concentration and hydrogen production rate by dark and semi-photo fermentation. Both H_2 and O_2 accumulated in the reactor headspace when a pure culture of *C. pseudolambica* L3 was tested. The O_2 level increased from 0% to 26% at 196th hour and to 38% at the end of reaction in the dark reactor, while the H_2 production rate initially increased from 2.5 to 22.3 ml H_2/gCDW/h in the first 72 hours, and then decreased to 0.65 ml H_2/gCDW/h. The low O_2 level of 3.6% allowed high H_2 production and 26 ml H_2/g cell dry weight was produced at the 24th hour. When O_2 level increased over 30%, H_2 productivity decreased nearly to zero both in dark and light phase showing again that [FeFe]-hydrogenase-like genes may exist and the results suggested that they were involved in the regulation of sensitivity to oxygen in strain *C. pseudolambica* L3 [23,24].

Fig. 6 describes the relationship between H_2 yield and both some anion / ion concentrations and VFAs in the liquid end products. NO_3^- anion and NH_4^+ ion concentrations were firstly compared (Fig. 6(a)) based on the analysis of effects of nitrogen sources. Concentration of NO_3^- anion was much lower in the dark condition than in the light/night condition and NH_4^+ ion concentrations were almost equal in these two reactors at the end of the reaction. It means much more N sources were reduced and more than 3 times H_2 production was obtained. On the other hand, NO_3^- anion as well as SO_4^{2-} anion in the dark reactor showed a lower value and a higher H_2 production than in light/night reactor maybe because more available hydrogen was consumed by sulphate and nitrate reducers and this led to a lower concentration of NO_3^- anion and SO_4^{2-} anion in the dark reactor (Fig. 6 (a) and (b)). Sulphate was reduced by molecular hydrogen to hydrogen sulfide. In a sense, some organisms "breathe" sulfate rather than oxygen in a form of anaerobic respiration. Nitrate was reduced to nitrite and more nitrite was obtained in the dark reactor (data not shown).

Dark-fermentation process involves aqueous metabolites during conversion of organic substrates to H_2 ((1)-(3)) [25]. However, the production of VFAs affects the buffering capacity that can inhibit the functioning of acidogenic yeasts. In this study (Fig. 6(c),(d)), a higher proportion of acetic acid along with butyric acid and propionic acid was observed in the controls. More acetic acid and H_2 production were obtained in dark phase than in natural light/night phase. Molar acetic acid to butyric acid ratio was 41 in dark reactor and 25 in the other reactor, respectively. The ratio of acetic acid yield to butyric acid yield (A/B ratio) was also considered as an indicator for evaluating the efficiency of hydrogen production during glucose fermentation with microorganisms [26]. A high molar acetate to butyrate ratio leads to more hydrogen production with acetate as a product (4molH_2/mol-acetate) than with butyrate (2molH_2/mol-butyrate) [27]. Besides, compared to the propionic acid pathway, the acetate or butyrate pathway is more favorable for H_2 production [28].

$$C_6H_{12}O_6{+}2H_2O \quad 2CH_3COOH{+}2CO_2{+}4H_2 \tag{1}$$

$$C_6H_{12}O_6{+}2H_2O \quad 2CH_3\,CH_2\,CH_2COOH{+}2CO_2{+}2H_2 \tag{2}$$

$$3C_6H_{12}O_6 \quad 4CH_3\,CH_2COOH{+}2CO_2{+}2H_2O \tag{3}$$

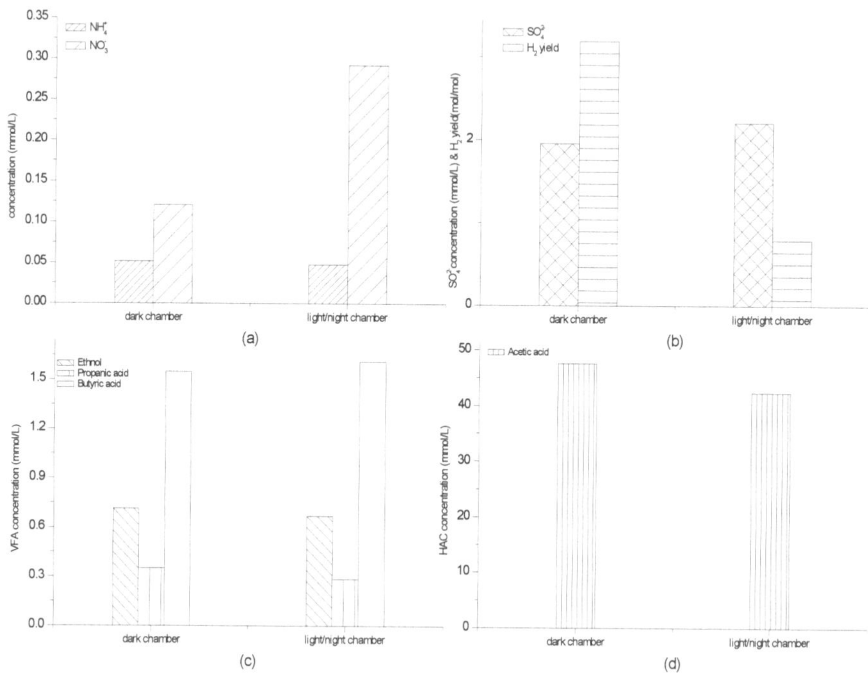

Figure 6: Profiles of liquid end products in dark and diurnal conditions. (a) NH_4^+ and NO_3^- concentration; (b) SO_4^{2-}concentration and H_2 yield; (c) VFAs concentration; (d) HAC concentration

4 Conclusions

A hydrogen-producing strain L3 was isolated and identified as *C. pseudolambica*. It could format biohydrogen using glucose and starch as substrates and ammonium sulfate as nitrogen source efficiently. Fe^{3+} ions were the most favorable metal element for cell growth and hydrogen production, especially under a dark environment. The maximum hydrogen yield of 3.19 mol H_2/mol glucose was obtained and [FeFe]-hydrogenase-like genes was supposed to be expressed in it. In addition, acetic acid and butyric acid other to propanoic acid were the main metabolite production profiles and a higher hydrogen production was obtained subsequently. Comparing with other biohydrogen microorganisms, *C. pseudolambica* L3 has the advantages of compatative higher hydrogen production ability with simple culture demand and lower economical cost.

Acknowledgment: This work was supported by the Beijing Municipal Natural Science Foundation (grant number 8112009) of China.

References

[1] Zhao X., Xing D., Fu N., Liu B., Ren N., Hydrogen production by the newly isolated Clostridium beijerinckii RZF-1108[J], Bioresour Technol., 2011102(18):8432-8436.

[2] Patel S.K.S., Kumar P., Singh M., Lee J.K., Kalia V.C.,. Integrative approach to produce hydrogen and polyhydroxybutyrate from biowaste using defined bacterial cultures[J], Bioresour Technol., 2015, 176:136-141.

[3] Mao C., Feng Y., Wang X., Ren G., Review on research achievements of biogas from anaerobic digestion [J], Renew Sustain Energy Rev., 2015, 45:540-555.

[4] Gou, C., Guo, J., Lian, J., Guo, Y., Jiang, Z., Yue, L., et al. Characteristics and kinetics of biohydrogen production with Ni2+ using hydrogen-producing bacteria [J], Int J Hydrogen Energy, 2015, 40(1):161-167.

[5] Zhang Z-P, Show K-Y, Tay J-H, Liang DT, Lee D-J, Jiang W-J. Effect of hydraulic retention time on biohydrogen production and anaerobic microbial community[J], Process Biochem., 2006,41(10):2118-2123.

[6] Azbar N., Çetinkaya Dokgöz F.T., Keskin T., Korkmaz K.S., Syed H.M., Continuous fermentative hydrogen production from cheese whey wastewater under thermophilic anaerobic conditions[J], Int J Hydrogen Energy, 2009,34(17):7441-7447.

[7] Wang M.Y., Olson B.H., Chang J.S., Relationship among growth parameters for Clostridium butyricum, hydA gene expression, and biohydrogen production in a sucrose-supplemented batch reactor[J], Appl Microbiol Biotechnol., 2008, 78(3):525-532.

[8] Melis A., Zhang L., Forestier M., Ghirardi M.L., Seibert M.. Sustained photobiological hydrogen gas production upon reversible inactivation of oxygen evolution in the green alga Chlamydomonas reinhardtii[J], Plant Physiol., 2000,122(1):127-136.

[9] Hall DO, Markov SA, Watanabe Y, Krishna Rao K. The potential applications of cyanobacterial photosynthesis for clean technologies[J], Photosynth Res., 1995,46(1-2):159-167.

[10] Wu TY, Hay JXW, Kong LB, Juan JC, Jahim JM. Recent advances in reuse of waste material as substrate to produce biohydrogen by purple non-sulfur (PNS) bacteria[J], Renew Sustain Energy Rev., 2012,16(5):3117-3122.

[11] Kumar, P., Sharma, R., Ray, S., Mehariya, S., Patel, S.K.S., Lee, J.K., et al. Dark fermentative bioconversion of glycerol to hydrogen by Bacillus thuringiensis[J], Bioresour Technol., 2015,182:383-388.

[12] Zagrodnik R, Laniecki M. An unexpected negative influence of light intensity on hydrogen production by dark fermentative bacteria Clostridium beijerinckii[J], Bioresour Technol., 2016,200:1039-1043.

[13] Fabiano B, Perego P. Thermodynamic study and optimization of hydrogen production by Enterobacter aerogenes[J], Int J Hydrogen Energy, 2002,27(2):149-156.

[14] Oh Y.K, Kim H.J, Park S, Kim M.S, Ryu D.D.Y., Metabolic-flux analysis of hydrogen production pathway in Citrobacter amalonaticus Y19[J], Int J Hydrogen Energy, 2008,33(5):1471-1482.
[15] Kumar N., Das D., Enhancement of hydrogen production by Enterobacter cloacae IIT-BT 08[J], Process Biochem., 2000,35(6):589-593.
[16] Kumar N., Das D., Continuous hydrogen production by immobilized Enterobacter cloacae IIT-BT 08 using lignocellulosic materials as solid matrices[J], Enzyme Microb Technol., 2001,29(4-5):280-287.
[17] Xing D., Ren N., Li Q., Lin M., Wang A., Zhao L., Ethanoligenens harbinense gen. nov., sp. nov., isolated from molasses wastewater[J], Int J Syst Evol Microbiol., 2006,56(4):755-760.
[18] Ren N., Cao G., Wang A., Lee D.J., Guo W., Zhu Y., Dark fermentation of xylose and glucose mix using isolated Thermoanaerobacterium ihermosaccharolyticum W16[J], Int J Hydrogen Energy., 2008,33(21):6124-6132.
[19] Kumar G., Lin C-Y., Bioconversion of de-oiled Jatropha Waste (DJW) to hydrogen and methane gas by anaerobic fermentation: Influence of substrate concentration, temperature and Ph[J], Int J Hydrogen Energy, 2013,38(1):63-72.
[20] Owen W.F., Stuckey D.C., Healy J.B., Young L.Y., McCarty P.L., Bioassay for monitoring biochemical methane potential and anaerobic toxicity[J], Water Res., 1979,13(6):485-492.
[21] Zhao, W., Fang, B.X., Niu, Y.J., Liu, Y.N., Liu, B., Peng, Q., et al., Nar1 deficiency results in shortened lifespan and sensitivity to paraquat that is rescued by increased expression of mitochondrial superoxide dismutase[J], Mech Ageing Dev., 2014,138(1):53-58.
[22] Winkler M., Esselborn J., Happe T., Molecular basis of [FeFe]-hydrogenase function an insight into the complex interplay between protein and catalytic cofactor[J], Biochim Biophys Acta - Bioenerg., 2013,1827(8-9):974-985.
[23] Ginger M.L., Fritz-Laylin L.K., Fulton C., Cande W.Z., Dawson S.C., Intermediary Metabolism in Protists: A Sequence-based View of Facultative Anaerobic Metabolism in Evolutionarily Diverse Eukaryotes[J], Protist., 2010,161(5):642-671.
[24] D'Adamo, S., Jinkerson, R.E., Boyd, E.S., Brown, S.L., Baxter, B.K., Peters, J.W., et al., Evolutionary and biotechnological implications of robust hydrogenase activity in halophilic strains of tetraselmis[J], PLoS One., 2014, 9(1).
[25] Venkata Mohan S., Lalit Babu V., Sarma P.N., Anaerobic biohydrogen production from dairy wastewater treatment in sequencing batch reactor (AnSBR): Effect of organic loading rate[J], Enzyme Microb Technol., 2007,41(4):506-515.
[26] Khanal S.K., Chen W.H., Li L., Sung S., Biological hydrogen production: Effects of pH and intermediate products[J], Int J Hydrogen Energy, 2004,29(11):1123-1131.
[27] Van Ginkel S.W., Logan B., Increased biological hydrogen production with reduced organic loading[J], Water Res., 2005,39(16):3819-3826.
[28] Ren N., Liu B., Ding J., Guo W., Cao G., Xie G., The effect of butyrate concentration on photo-hydrogen production from acetate by Rhodopseudomonas faecalis RLD-53[J], Int J Hydrogen Energy, 2008,33(21):5981-5985.

Rongjuan Yang*, Chuxue Wu, Yuhui Ge, Cheng Fang, Decai Yang

Cancer Trends Predication Based on an Intelligent Method

Abstract: Cancer incidence is a complicated dynamic system affected by many factors. Predicting cancer trends can help evaluate the progress in cancer control while reinforcing prevention activities. To forecast cancer trends, one mathematical model to analyze the development of cancer incidence is proposed. Based on the analysis of this model, one new intelligent method for predicting cancer trends has been proposed. For this new intelligent method, one new evolutionary neural network, combined with the modified back propagation algorithm and the black hole algorithm, was applied. Finally, long term monitoring of data on cancer trends from Canada and China were used to verify the good performance of this new intelligent method.

Keywords: cancer incidence, cancer trends, predication, evolutionary neural network, intelligent method

1 Introduction

Nowadays, cancer has become a leading cause of the death throughout the world. In China, it is the leading cause of death in urban areas and the second most common cause in rural areas for many years [1]. Many important factors, such as the ageing population, severe environmental pollution, uncontrolled chronic infection, and adoption of westernized lifestyles, cancer in China is expected to become more and more severe [2]. Moreover, in Canada, at the beginning of 2007, nearly 750 000 Canadians had a diagnosis of cancer more than in the previous 10 years [3]. Therefore, cancer has become the enemy of human health. To control the cancer incidence, it is very important to predicate the cancer trends based on the long time monitoring of cancer data.

***Corresponding author: Rongjuan Yang,** Hospital of Wuhan Polytechnic University, Wuhan, Hubei, 430023 P. R. China, E-mail: xiao12yang@163.com
Chuxue Wu, School of Optical-Electrical and Computer Engineering, University of Shanghai for Science & Technology, Shanghai, 200093 P. R. China
Yuhui Ge, Business School, University of Shanghai for Science & Technology, Shanghai, 200093 P. R. China
Cheng Fang, College of Health Science & Nursing, Wuhan Polytechnic University, Wuhan, Hubei, 430023 P. R. China
Decai Yang, Hubei University of Chinese Medicine, Wuhan, Hubei, 430065 P. R. China

To predict cancer trends, some basic laws hidden in the historical data of cancer incidence such as some mathematic functions, must be recognized and analyzed [4-5]. Because the artificial neural network (ANN) can draw its law from very complicated data, and can approximate very complicated function, then the artificial neural network can be used to model the non-linear dynamic system [6]. Generally, modeling the non-linear dynamic system is to model the time series of this system. The ANN can solve this problem very well [7]. Therefore, the ANN can be used to predicate cancer trends. To improve the ANN, in this study, one new intelligent method using one new evolutionary neural network (ENN) is applied. At last, some long time monitoring of cancer trends data from Canada and China were used to verify the good performance of this new intelligent method.

2 Mathematic Model of Cancer Incidence

As the cancer incidence system is a complicated and dynamic system [4-5], it can be described mathematically as follows.

The evolvement of a dynamic system can be analyzed using some differential equations [8]. In this study, the cancer incidence is supposed to be described by one q-th order ordinary differential equation, as.

$$\frac{d^p x}{dx^p} = f(x, x', \cdots, x^{(p-1)}) \tag{1}$$

If the solution for above equation is one state variable of this system, then the time series of this variable can be used to model the dynamic system. In this study, the follow third order ordinary differential equation is applied.

$$\frac{d^2 x}{dt^2} = f(x, \frac{dx}{dt}) \tag{2}$$

If a time series $x(j)$ $j = 1 \sim n$ with step h, is the solution of the equation (2), then the numerical solution of the equation (2) can be used to model the system, which is,

$$\left.\frac{d^2 x}{dt^2}\right|_j = \frac{x(j+1) - 2x(j) + x(j-1)}{h^2} + o(h^2) \tag{3}$$

$$\left.\frac{dx}{dt}\right|_j = \frac{x(j) - x(j-1)}{h} + o(h) \tag{4}$$

Thus, the time series can be described as,

$$x(j+1) = h^2 f[x(j) \frac{x(j) - x(j-1)}{h}] + 2x(j) - x(j-1) + o(h^3) \tag{5}$$

The function (5) can be generalized to p-th order by retrieving its high order minim. Thus, the time series can be described as,

$$x(j+1) = F[x(j) \; x(j-1) \; h] \tag{6}$$

Therefore, we can see that, the above function can be used to describe the dynamic evolvement model of cancer incidence. And the late state of this system can be described by some of its prophase states.

In the traditional method, the function relationship F is supposed as a linear function. But in the cancer incidence system, the F should be a very complicated non-linear function. Thus, the above function is always very complicated and its description cannot be confirmed. To this problem, the traditional method cannot solve it. For the ANN, the information is stored in linkage of network hidden, and cannot be described with one function. Moreover, the neural network has the adaptivity for a dynamic process [9]. Thus, the ANN can be used to approximate the function relationship F of equation (6) very well.

3 Evolutionary Neural Network for Cancer Trends Predication

From the above analysis, the number of input neuron for ANN can be confirmed by the dynamic system, but the cancer incidence system described by n-th order differential equation is an uncertain system. Therefore, the lingering time step of this time series for the ANN cannot be obtained. However, it is a very important parameter to construct ANN samples, as it can seriously affect the predication results [9]. Nowadays, there is no good method to determine this parameter. Normally, in general applications, it is determined by the experiment.

Therefore, to model the cancer incidence system by the ANN, the most important work is to determine the construction of the ANN. Because the selection of the construction of the ANN is a very typical optimization step, the global optimization-evolutionary algorithm can solve this problem very well [10]. And this kind of ANN is called ENN. Nowadays, the ENN has been used in time serials data forecasting field [11].

In this study, the neural numbers of hidden layer and input layer are determined by the evolutionary algorithm, and one new ENN is proposed. In this new ENN, the construction of ANN is determined by the black hole algorithm [12] and the weights are determined by a modified back propagation algorithm. In this study, the three layers ANN is used. Therefore, in this ENN, there are four parameters to be evolved by the black hole algorithm, which are the numbers of the input neuron and the hidden layer neuron, iterating step η and inertia parameter α. Moreover, in this study, the indirect coding method and full-linking network are used.

3.1 Black Hole Algorithm

One new nature-inspired metaheuristic algorithm inspired from the black hole phenomenon, is the black hole algorithm (BHA) which was firstly proposed by

Hatamlou [10] in 2013. Studies show that, for BHA, the controlling parameters are very few and the structure is very simple. However, the performance of the BHA is very good. Thus, nowadays, the BHA has been used in engineering problems [13]. In this ENN, to obtain the good ENN, the BHA is used.

The detailed process for the BHA is as following.

1. *Creation the initial stars:* Generally, in the search range of the problem, the initial stars (candidate solutions) are created randomly. In this algorithm, one star is its representation. In other words, for the search range of a n-dimension space, the star is a n-dimension vector.
2. *Fitness value computation:* The fitness value is to evaluate the star. Moreover, the progression of the algorithm is controlled by the fitness value, which is its criteria. In this method, the transformation of the objective function for one star is its fitness function.
3. *Black hole selection:* The star whose fitness value is the best is selected as the black hole.

The stars that surround the black hole will be absorbed by the black hole. Thus, when the first black hole is initialized, all the stars that start to move towards the black hole and the stars around it will be absorbed.

4. *Stars movement:* Because the black hole has the ability to absorb the stars that surround it, thus, all stars must move towards the black hole and change their locations. This process is as follows,

$$x_i(t+1) = x_i(t) + rand \times (x_{BH} - x_i(t). \tag{7}$$

where n is the stars number; *rand* is a random number whose interval is [0, 1]; x_{BH} is the location of the black hole; $x_i(t)$ and $x_i(t+1)$ are the locations of the i-th star at iterations t and $(t + 1)$ respectively; $i = 1, 2, 3, \ldots, n$.

5. *Stars replacement:* For BHA, to select the more optimal star, the probability to cross over the event horizon for the black hole by the moving stars is applied. The radius of the event horizon is calculated as follow,

$$R = \frac{f_{BH}}{\sum_{i=1}^{n} f_i}. \tag{8}$$

where f_{BH} is the fitness values of the black hole and f_i is the fitness value of the i-th star.

6. *Update black hole:* In the moving process of stars, when a star reaches one location whose cost is the lowest cost, that is, the fitness value of this star is better than that of the black hole. Thus, this star will replace the black hole. This process is called update black hole.

   After all the stars have been moved, the next iteration will begin.
7. *Termination condition:* In this study, when the number of iterations performed without replacing the black hole reaches five, the algorithm process will stop. In other words, if the black hole is not replaced for five iterations, the algorithm will stop. Moreover, to avoid infinite iteration of the algorithm process, in this study, the maximum number of iterations must be given.

Based on above analysis, the basic flow chart for the BHA can be determined, shown in Fig. 1.

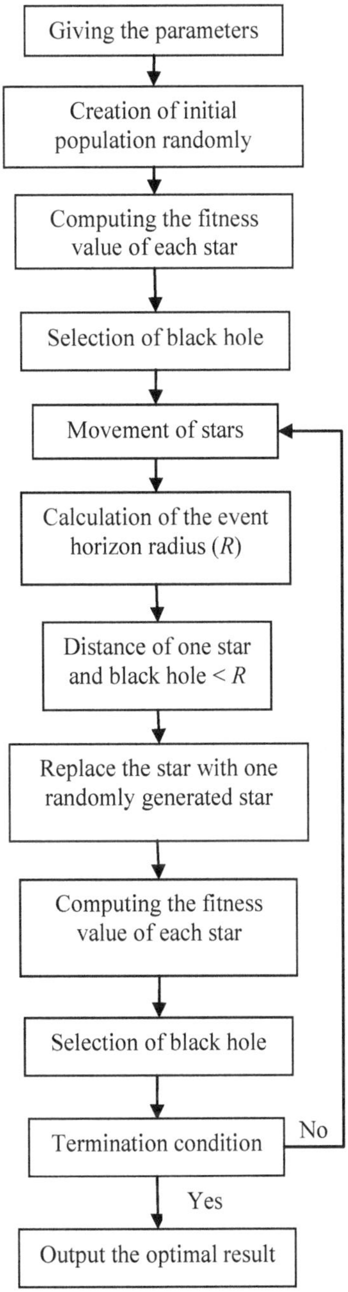

Figure 1: Flow chart of BHA

3.2 Evolutionary Neural Network

The details of ENN are as follows.

1) Firstly, the search ranges for the four optimization parameters, which are numbers for the input and layer neurons, iterating step η and inertia parameter α, must be given. Moreover, the controlling parameters for ENN, including maximum number of iterations, star number in one population, output neuron number, criteria for the iterating stop and iterating error of the modified back propagation algorithm, must be given too.
2) In this ENN, two random numbers for neuron of the input and hidden layers can determine the network construction. And two random values for parameters η and α can determine one modified back propagation algorithm. Using this network construction and this modified back propagation algorithm, one initial star can be created.
3) The fitness value for one star is computed by the following steps.
   a) According to the number of input neuron, the whole time series of cancer trends data is divided to construct the learning samples for the ENN.
   b) To improve the performance of ENN, the learning samples are divided into two parts, which are training samples and testing samples. The training samples are used to determine the non-linear mapping of the ANN, and the testing samples are used to test the generalization of the ANN.
   c) The initial linking weights for one star are generated randomly.
   d) For the modified back propagation algorithm, the iterating step is taken as $j = 1$.
   e) For one individual, its square error $E(j)$ trained by testing samples is computed, and taken as the minimum error, $\min E = E(j)$. And then, the algorithm process returns to step (3).
   f) For one individual, if the square error trained by training samples is less than the iterating error criteria for modified back propagation algorithm, its fitness value is still to be $\min E$. And then, the process returns to step (3) too.
   g) Using the modified back propagation algorithm, the linking weights for the individual stars are adjusted.
   h) The iterating step increased by one, that is, $j = j + 1$, then, the process returns to step e.
   i) If the iterating step reaches the given iterating stop criteria for modified back propagation algorithm, the fitness value of the individual star equals to $\min E$. And then, the process returns to step (3) too.
4) The black hole is selected.
5) All the individual stars in the population are moved.
6) The event horizon radius of the individual stars are computed. And the sucked stars are replaced.
7) Using the fitness computation method shown in step (3), the fitness values for all individual stars are computed.

8) The black hole is selected again.
9) The number of iterations increases 1. If the number of iterations reaches its stop criteria, the algorithm process will stop. Therefore, the black hole at this time is the best ANN for prediction the time series of cancer trends data.

Using the best searched ANN for predicting the time series of cancer trends data, the real time series of cancer trends data are applied, and the forecasting data can be obtained.

4 Real Example Application

Using the obtained ENN, some long term monitoring of cancer data from Canada and China have been used to predict cancer trends.

Firstly, the data of four main cancers in Canada, which are prostate cancer for males, lung cancer for males, colorectal cancer for males, and breast cancer for females, were applied. In this study, 21 sets of data from 1970-1990 for the age-standardized incidence rate were used [3], which are shown in Figures 2-5. The detailed description of long term monitoring of cancer data can be found in the reference [3].

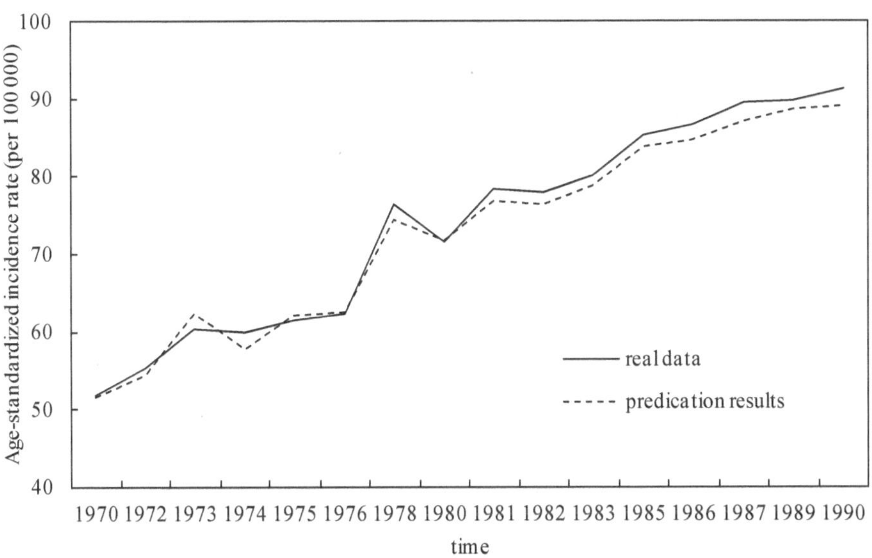

Figure 2: Data of prostate cancer for males

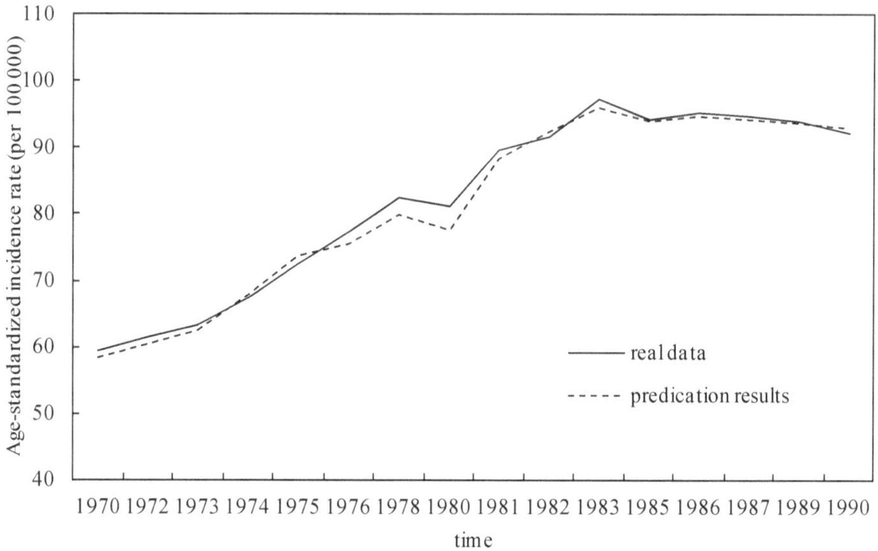

Figure 3: Data of lung cancer for males

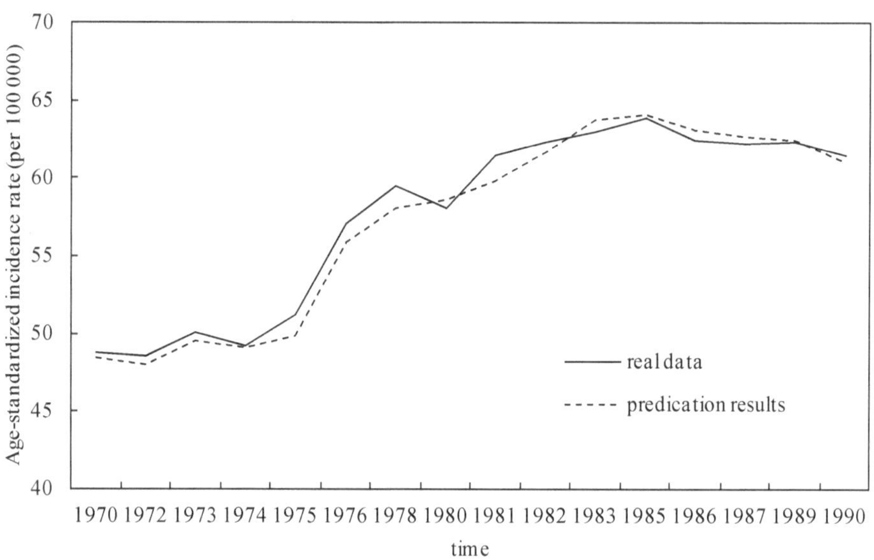

Figure 4: Data of colorectal cancer for males

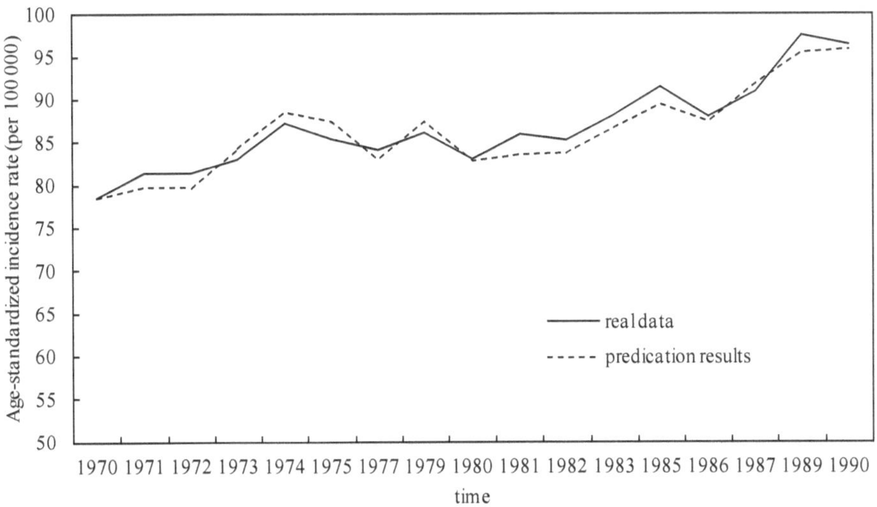

Figure 5: Data of breast cancer for females

In this study, to construct the ENN model, the former 19 sets of data were used. Moreover, to verify the predication, the last 2 data sets were applied. In the 19 data sets, the former 14 sets of data were selected to establish the training samples, and the last 5 data sets were selected to establish the testing samples. For four data sets, the controlling parameters of ENN were as follows,

The maximum number of iterations was 200. The star number for one population was 50. The output neuron was 1. All the criteria of the iterating stop and iterating error for the modified back propagation algorithm were 10e-5.

After computation, the best construction of ANN for four data sets can be obtained, which are 6-11-1 for the data set of prostate cancer, 7-15-1 for the data set of lung cancer in males, 6-13-1 for the data set of colorectal cancer in males, and 8-12-1 for the data set of breast cancer in females. Using those ANNs, the predication results for four data sets of cancer trends are shown in Fig. 2-5.

From the above four figures, it can be found that, the predication results of new intelligent method are comparable with the real monitoring data, thus, the cancer trends can be predicted by the new intelligent method very well.

Finally, the data from two main types of cancers in China, which are lung cancer for males and breast cancer for females, were applied. In this study, 12 sets of data from 2000-2011 for the incidence rate were used [14], which are shown in Figures 6 and 7. The detailed description of the long term monitoring of cancer data can be found in the reference [14].

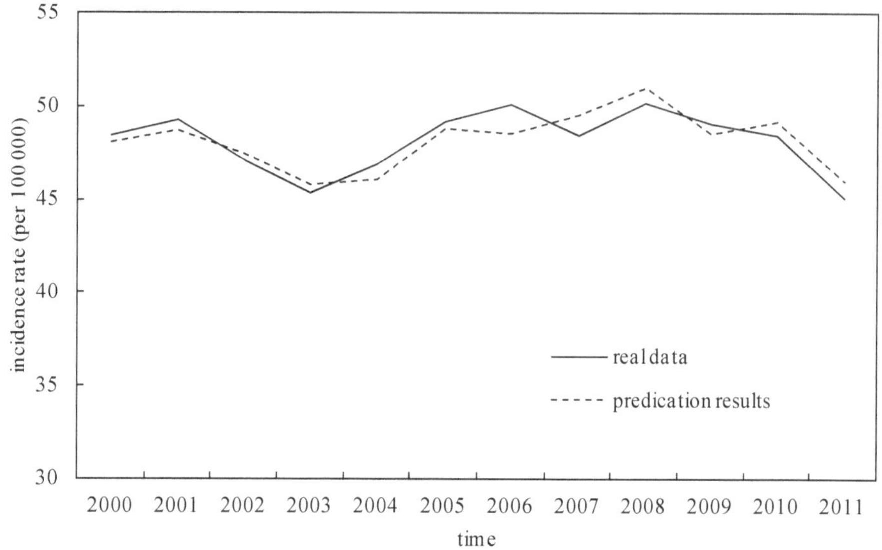

Figure 6: Data of lung cancer for males

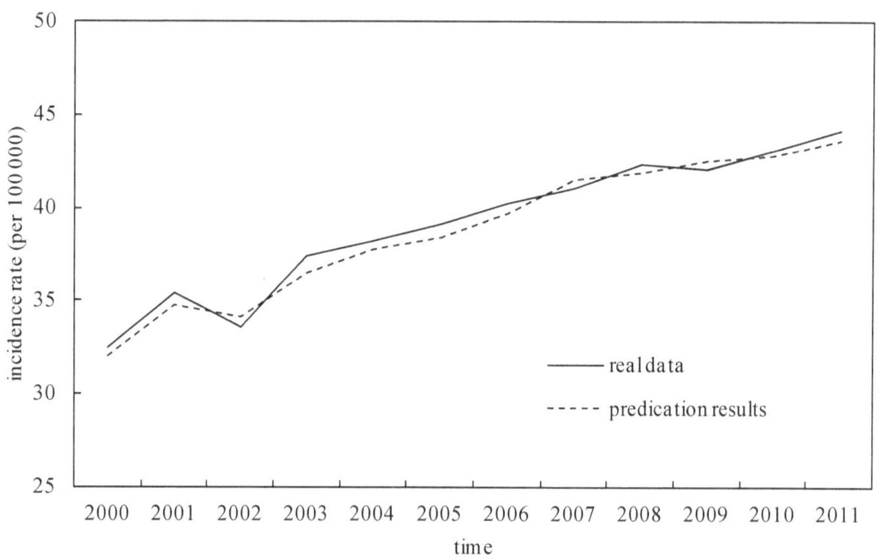

Figure 7: Data of breast cancer for females

In this study, to construct the ENN model, the former 9 data sets were used. Moreover, to verify the predication, the last 2 data sets were used. In the 9 data sets, the former 7 data sets were selected to establish training samples, and the last 2 data sets were selected to establish testing samples. For two data sets, the controlling parameters of ENN were as follows,

The maximum number of iterations was 200. The star number for one population was 50. The output neuron was 1. All the criteria of the iterating stop and iterating error for the modified back propagation algorithm were 10e-5.

After computation, the best construction of ANN for two data sets were obtained, which are 4-15-1 for data set of lung cancer for males and 4-12-1 for data set of breast cancer for females. Using those ANNs, the predication results for two cancer trends data set are shown in Fig. 6 and 7.

From the above two figures, the predication results of the new intelligent method correlate with the real monitoring data well, thus, the cancer trends can be predicted by the new intelligent method very well.

5 Conclusions

Predication of cancer trends can help to evaluate the progress in cancer control while reinforcing prevention activities. In this study, the change law of cancer incidence is analyzed by the mathematical method, and one new intelligent method to predict the cancer incidence trends based on new ENN is proposed. Finally, using six long term monitoring of cancer trends data from Canada and China, this new intelligent method was verified. The predication results show that, the new intelligent algorithm can be applied to predict the real cancer incidence trends very well.

References

[1] Chen Z., The Third National Retrospective Sampling Survey on Causes of Death [M], Beijing: Peking Union Medical College Press, 2008: 5-30.

[2] Goss P.E., Strasser-Weippl K., Lee-Bychkovsky B.L., Fan L., Li J.J., Chavarri-Guerra Y., et al, Challenges to effective cancer control in China, India, and Russia [J], Lancet Oncol., 2014, 15: 489-538.

[3] Canadian Cancer Society's Steering Committee on Cancer Statistics. Canadian Cancer Statistics 2011. Toronto (ON): Canadian Cancer Society, 2011: 25-54.

[4] Zeng H.M., Zheng R.S., Guo Y.M., Zhang S.W., Zou X.N., Wang N., et al, Cancer survival in China, 2003–2005: A population-based study [J], Int. J. Cancer, 2015,136(8): 1921-1930.

[5] Kachuri L., De P., Ellison L.F., Semenciw R.,, Cancer incidence, mortality and survival trends in Canada, 1970–2007 [J], Chronic Dis. Inj. Can., 2013, 33: 69-80.

[6] Luo F.L., Unbehauen R., Applied Neural Networks for Signal Processing [M], New York: Cambridge University Press, 1997: 121-130.

[7] Gao W., Modeling Methods Analysis for Non-linear Time Series [J], J. Tsinghua Univer., 2000, 37: 6-10.
[8] Beltrami E.D., Mathematics for Dynamic Modeling [M], New York: Academic Press, 1998: 115-139.
[9] Jacek Z.M., Introduction to Artificial Neural Systems [M], St. Paul: West Publishing Company, 1992: 57-60.
[10] Jin F., Neural Computing Intelligence Foundation-Principles and Method [M], Chengdu: Xi'nan Jiaotong University Press, 2000: 158-165.
[11] Gao W., New Neural Network Based on Ant Colony Algorithm for Financial Data Forecasting [C], New York: IEEE Press, 2008: 1437-1440.
[12] Hatamlou A., Black hole: a new heuristic optimization approach for data clustering [J], Inform. Sciences, 2013, 222: 175-184.
[13] Gao W., Ge M.M., Chen D.L., Wang X., Back Analysis for Rock Model Surrounding Underground Roadways in Coal Mine Based on Black Hole Algorithm [J], Eng. Comput., 2016, 32(4): 675-689.
[14] Chen W.Q., Zheng R.S., Baade P.D., Zhang S.W., Zeng H.W., Bray F., et al, Cancer Statistics in China, 2015 [J], CA: Cancer J. Clin. 2016, 66(2): 115-132.

Shen Wei*, Chen Juan

Effect of Lead on Reproductive System and Levels of Reproductive Hormones in SD Rats' Gestation Period

Abstract: Objective: to study the toxicity damage mechanism of the reproductive system of SD rats induced by lead exposure. Methods: Beginning from the first day after SD rats' conception, the rats in the low lead group and high lead group were provided with an unlimited amount of distilled water with 100 and 200 mg/L of lead acetate. The rats in the control group were provided with the same amount of distilled water. At term (20th day), the blood lead of pregnant rats was detected by atom absorption spectrum, and radioimmunoassay (RIA) was used to determine the levels of follicle-stimulating hormone (FSH), luteotropic hormone (LH), prolactin (PRL), estradiol (E2) and progesterone (P). The fine structure of uteruses and ovaries was observed under a light microscope. Results: Compared with the control group, gestational lead exposure significantly increased the blood lead levels of pregnant rats ($P < 0.05$), while reducing FSH and E2 levels ($P < 0.05$), increasing LH, PRL and P levels ($P < 0.05$), and resulted in varying degrees of pathological changes to the uterine tissue. Conclusion: Lead causes different degrees of damage to the reproductive system of SD rats during pregnancy and interferes with its normal function.

Keywords: SD rat, lead, reproductive system.

1 Introduction

Lead (Pb) was a common poison in the daily environment. With the development of industry, there was an increasing chance of contracting lead poisoning in life. The polyaffinity and accumulation of lead could cause disorders of the multi-organ and multi-system, including the female reproductive system. For pregnant women, studies have shown that blood lead levels exceeding the normal level could increase the incidence of some diseases such as premature rupture of membranes, premature birth, hypertensive disorder complicating pregnancy, abortion, embryonic development stops, and so on. The higher the level of blood lead in the mother, the risk of premature rupture of membranes in late trimester of pregnancy was greater, and higher doses of lead exposure could also result in embryo morbidity and stillbirth [1-4]. But its pathogenic mechanism and histopathology change still needed to be

***Corresponding author: Shen Wei,** Department of Obstetrics and Gynecology, The Forth Hospital of Hebei Medical University, Shijiazhuang, China, E-mail: zxhpjyscfl@163.com
Chen Juan, Health Examination Center, The Forth Hospital of Hebei Medical University, Shijiazhuang, China

further explored. In this experiment, a lead-exposure animal model of pregnancy was established. The content of lead in the peripheral blood and the level of reproductive hormones in serum were determined, and the pathological morphology of uterus and ovary of pregnant SD rats were analyzed. It provided a theoretical basis for the pathogenesis and treatment of lead exposure during pregnancy.

2 Material and Method

2.1 Main Material

Lead acetate [$Pb(Ac)_2$, Analytical reagent] was purchased from Guangzhou Chemical Reagent Factory. Iodine [125^I]- human follicle-stimulating hormone (FSH), iodine [125^I]- human luteotropic hormone (LH), prolactin (PRL), iodine [125^I]- estradiol (E2), and iodine [125^I]- human serum progesterone(P) were purchased from Beijing-Dongya Biotechnology Research Institute. WFX-120 atomic spectrophotometer was obtained from Japan Hitachi; SN-684 type γ RIA counter was obtained from Shanghai Institute of Nuclear Research; UFX-II microscope was obtained from Japan Olympus. Adult SD rats were purchased from Experimental Animal Center of Hebei Medical University (Certificate of conformity: DK0503-0031).

2.2 Experiment Methods

1. *Lead exposure model of pregnant rat:* At 8:00pm, female rats were mated with male rats at 2:1. At 8:00am, the vaginal pessary of female rats were checked, and the day appeared vaginal pessary was taken as the first day of pregnancy. SD rats were randomly divided into 3 groups, including control group, the low lead group and the high lead group. From the first day of conception, the pregnant rats in the low lead and high lead groups were given distilled water, containing 100, 200 mg/L lead acetate for 20 days and the control group was distilled water.
2. *Sample drawing:* All the samples were collected by sterile test tubes. The rats in each group were, at the 20th day, anesthetized by 2% pentobarbital sodium for 3-5 mg per 100 g weight through intraperitoneal injection. Blood was taken from the heart to determine the level of lead and reproductive hormone, and the uterus and ovary tissue were obtained for morphological analysis.
3. *Determination of blood lead and reproductive hormones:* On the 20th day of pregnancy, 2% pentobarbital sodium (30-50 mg/kg) was used to intraperitoneal anesthesia and the blood was obtained from the heart. Follicle-stimulating hormone (FSH), luteotropic hormone (LH), prolactin (PRL), estradiol (E2) and progesterone (P) were measured by radioimmunoassay (RIA), and the level of blood lead was measured by WFX- 120.

4. *Histopathological examination:* The heart blood was taken, and at the same time the uterus and ovary tissue were taken out. The tissue was fixed with 10% formalin, embedded in paraffin and stained with HE method and analyzed by OLYMPUS light microscope.

2.3 Statistical Methods

SPSS12.0 software was used to complete the statistics, and test data was expressed as ($\bar{x} \pm s$). More than three groups were compared by variance analysis, and LSD method was used for further comparison between the two groups. P <0.05 was statistically significant

3 Results

3.1 The General Situation of SD Rats in Each Group

During the experiment, the diet and drinking water were normal, and no abnormality was observed, such as the following reactions: indifference, abnormal excitement, less eating and less moving, and fur dry, yellowing or shedding. The animal model requirements were ensured.

3.2 Lead Levels of Pregnant Peripheral Blood

The blood lead levels of pregnant rats in the control group, the low lead group and the high lead group were (60.32 ± 4.72), (105.70 ± 9.83) and (122.16 ± 9.38) μg/L respectively. Compared with the control group, the blood lead levels of pregnant rats in the low lead group and the high lead group increased significantly respectively, and differences were statistically significant (P <0.05).The blood lead levels in the high lead group were significantly higher than that in the low lead group (P <0.05).

3.3 Cell Morphological Observation of Reproductive System Uterus and Ovary Tissue in Pregnant Rats

Compared with the control group, the endometrial tissue showed a certain degree of loose cytoplasmic, cell edema, vacuolar degeneration, and other changes in the lead group, under the light microscope. But the changes of ovarian tissue in the lead group were not obvious under light microscope, compared with the control group, as shown in Fig. 1.

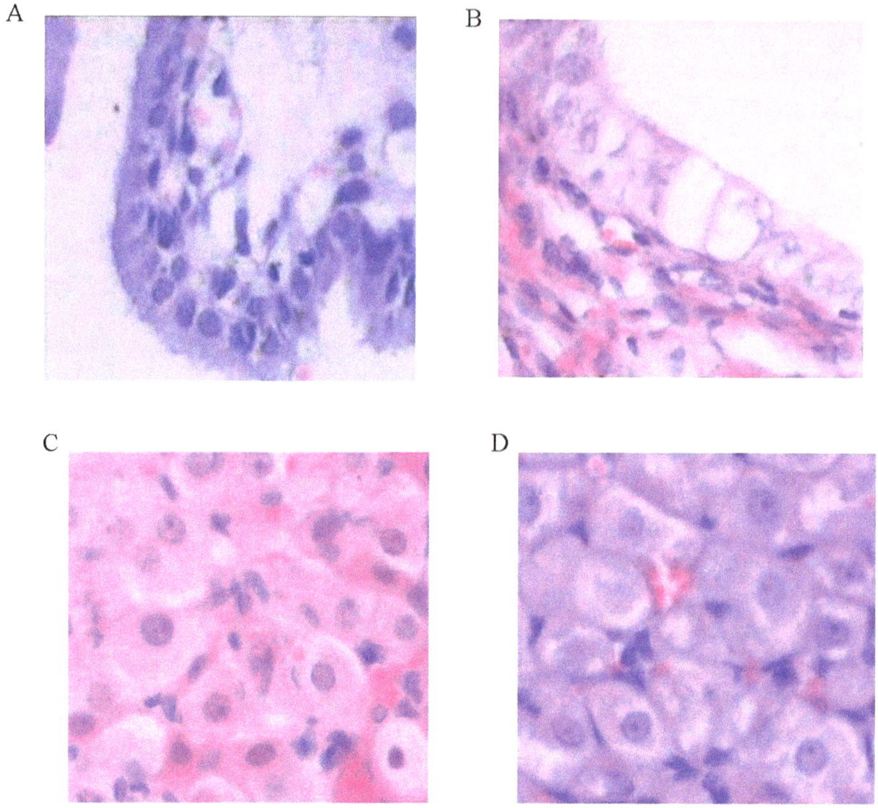

Figure 1: Cell morphology of uterine and ovarian tissue of reproductive system of the pregnant rats (HE × 400). (A) Uterine tissue in the control group; (B) Uterine tissue in the high lead group; (C) Ovarian tissue in the control group; (D) Ovarian tissue in the high lead group.

3.4 Reproductive Hormone Levels in Pregnant Rats

Compared with the control group, FSH and E2 levels decreased in the low lead group and the high lead group, while LH, PRL, P levels increased, and in addition to LH in the low lead group, differences were statistically significant ($P < 0.05$). Compared with the low lead group, FSH and E2 levels in the high lead group decreased, while LH, PRL and P levels increased, and except E2, differences were statistically significant ($P < 0.05$), as shown in Fig. 2.

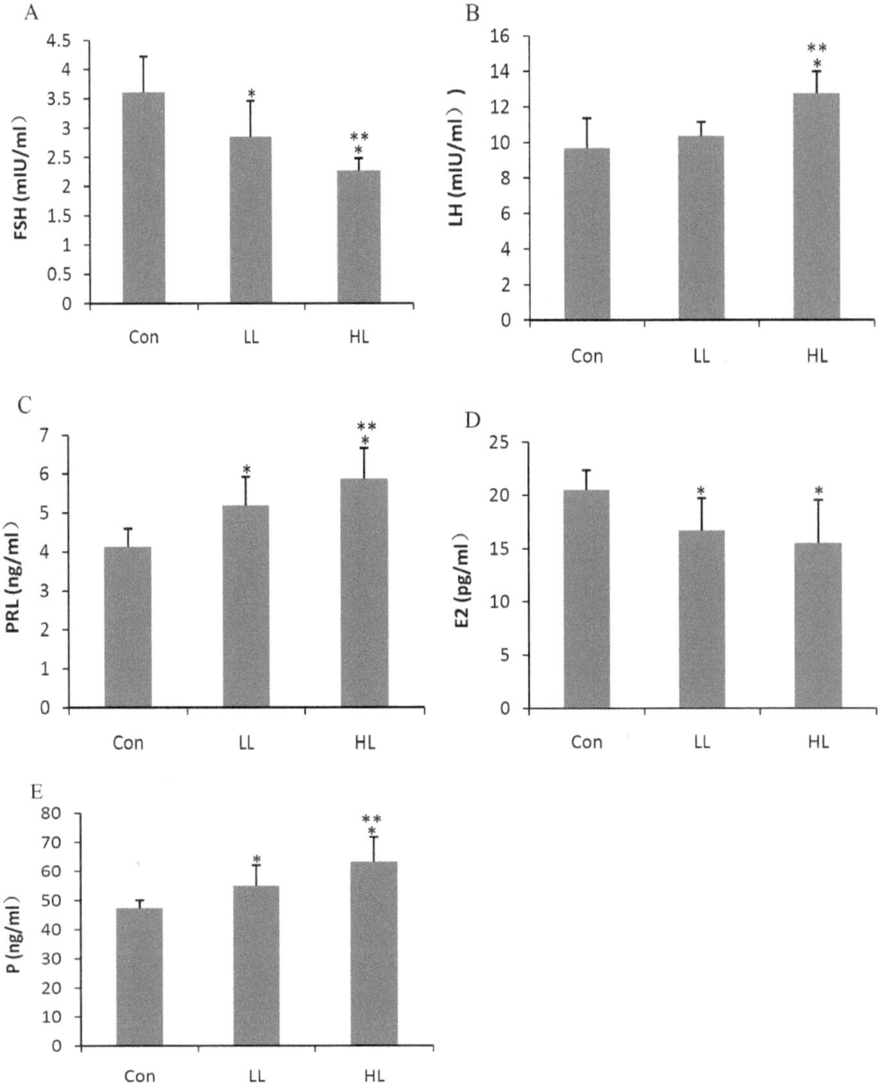

Figure 2: Reproductive hormone levels of serum of pregnant rats. (A) FSH; (B) LH; (C) PRL; (D) E2; (E) P. Con for the control group; LL for the low lead group; HL for the high lead group. *$P<0.05$, vs. Con; **$P<0.05$, vs. LL.

4 Discussion

In the early stage of pregnancy, women have a significantly increased sensitivity to the heavy metal lead, but blood lead levels of pregnant women in early pregnancy were generally lower than that of normal women. With the increase of gestation age, the blood lead levels of pregnant women increased gradually, and reached its peak in late pregnancy [5]. Therefore the late trimester of pregnancy, namely the 20th day after conception, was chosen as the terminal time to study the effect of the heavy metal lead on the pregnancy process and endocrine system.

When the human body is exposed to the heavy metal lead, its polyaffinity can cause harm to many organs and their physiological functions, including the female reproductive system [6]. Lead causes reproductive toxicity, embryo toxicity and teratogenic effect. Pregnant women with a high lead level had the following characteristics [7]: (1) Most of them had allotriophagy, (2) Nonspecific symptoms increased, such as discomfort, anemia, and basophilic cells increased in peripheral blood. (3) The blood lead levels of neonates were usually higher than their mother. Some studies found that lead could affect estradiol with the combination of human endometrial cells and muscle cell cytoplasm, which could hinder implantation, namely [8]: (1) Lead could induce oxygen free radicals, which are the key factors in initiating lipid peroxidation, these as well as the end products of lipid peroxidation of blood, liver, brain and other tissues, such as malondialdehyde (MDA), accumulate in the body, thereby leading to oxidative damage and thus irreversible damage of membrane, protein and DNA. (2) Lead could be related to thiol, which was contained by some important enzymes of the bodies antioxidant system, such as superoxide dismutase (SOD), glutathione peroxidase (GSH-Px), 6- glucose phosphate dehydrogenase (6-GPD). The combination of lead and these enzymes decreased their antioxidant capacity, thereby enhancing the harmful effect of lipid peroxidation, and leading to oxidative damage of the body. (3) Lead could inhibit the activity of Na + -K + -ATP enzyme to a certain extent.

In this study, through determining the blood lead levels of SD rats and analyzing the uterine and ovarian tissue in the experimental group and control group, it was found that with the increase of blood lead level, the pathological changes of the rats' genitals appeared to have different degrees of pathological changes and with the increase of lead content, the above phenomenon also increased. It could be inferred that the toxic effects of lead could affect the normal metabolism of the cells, making its normal function during pregnancy be disturbed. This was consistent with some studies in recent years [9]. The study found that pregnant women with high blood lead were more at risk of gestational hypertension, fetal malformations, fetal distress, abortion, oligohydramnios, premature birth, stillbirth, low birth weight children and disabled children, and an increase of blood lead per 50 μg/L, had an abortion rate increase of 1.8%.

In addition to acting on the internal genitalia of pregnant women, lead also could interfere with the normal metabolism of the reproductive endocrine system. During pregnancy, variation tendency of some reproductive hormones, such as follicle-stimulating hormone (FSH), luteotropic hormone (LH), prolactin (PRL), estradiol (E2) and progesterone (P), were as follows: (1) FSH and LH significantly decreased, because the inhibition effect of the negative feedback of hypothalamus–pituitary, resulted from a large increase of estrogen and progestin, which were secreted by the corpus luteum in early pregnancy and placenta in the following formation. (2) PRL during pregnancy progressively increased in the process. The study found that the PRL level of pregnant women with polyhydramnios, oligohydramnios, fetal intrauterine growth retardation and hypertensive disorder, complicated the already abnormal pregnancy, but the specific mechanism was unclear. (3) The E2 and P values gradually increased, and the ratio of E2/P increased significantly in late pregnancy [9]. Its significance was to promote cervical ripening and induce uterine contraction, so as to start labor and childbirth.

This study found that compared with control, in lead group the levels of FSH and E2 decreased, and the levels of LH, PRL and P increased significantly. The trend of these hormones were similar to the findings by Zhang SN et al. [10], but the statistical results of E2, FSH were not consistent, which could be related to the lead exposure dose, animal species, reproductive cycle and stage, and other factors. Govoni S et al had confirmed that the increase in PRL level following lead exposure in SD rats was related to the decrease in the number of dopamine receptors in the hypothalamus and pituitary of the pregnant rat induced by lead [11]. Through inhibiting the PRL, the dopamine adjusted hypothalamus and affected pituitary function, therefore the increase of LH could be related to the release level of norepinephrine induced by lead. Decrease of E2/P ratio could weaken uterine myofibril contraction, thereby resulting in an increase probability of abnormal labor. Liu W et al. found that abnormal births of women who work from home in an environment with high concentrations of lead, increased significantly, including a 13.8% increase in natural abortion rate, and a 15.9% increase in premature birth rate, and E2/P change induced by lead could be one of the factors that caused this phenomenon [12].

In summary, lead exposure during pregnancy could damage the reproductive system of SD rats, and affect the reproductive organs and endocrine system function, and after the toxic effect occurred, it was difficult to reverse. Therefore it is essential to reduce lead exposure during pregnancy

References

[1] Fadrowski J.J., Abraham A.G., Navas-Acien A., Guallar E., Weaver V.M., Furth S.L., Blood lead level and measured glomerular filtration rate in children with chronic kidney disease [J], Environ. Health Perspect., 2013, 121(8): 965–970.

[2] Taylor C.M., Golding J., Hibbeln J., Emond A.M., Environmental factors predicting blood lead levels in pregnant women in the UK: the ALSPAC study [J], PLoS One, 2013, 8(9): e72371.

[3] Ivanova L., Pentieva K., Tsachev K., The lead level in the blood of pregnant women from the city of Sofia [J], Akush Ginekol (Sofiia), 1997, 36(3): 7-9.

[4] Torres-Sánchez L.E., Berkowitz G., López-Carrillo L., Torres-Arreola L., Ríos C., López-Cervantes M., Intrauterine lead exposure and preterm birth [J], Environ. Res., 1999, 81(4): 297-301.

[5] Schell L.M., Czerwinski S., Stark A.D., Parsons P.J., Gomez M., Samelson R., Variation in blood lead and hematocrit levels during pregnancy in a socioeconomically disadvantaged population [J], Arch. Environ. Health, 2000, 55(2): 134-140.
[6] Hertz-Picciotto I., The evidence that lead increases the risk for spontaneous abortion [J], Am. J. Ind. Med., 2000, 38(3): 300-309.
[7] Shannon M., Severe lead poisoning in pregnancy [J], Ambul. Pediatr., 2003, 3(1): 37-39.
[8] Lawton L.J., Donaldson W.E., Lead-induced tissue fatty acid alterations and lipid peroxidation [J], Biol. Trace Elem. Res., 1991, 28(2): 83-97.
[9] Sowers M., Jannausch M., Scholl T., Li W., Kemp F.W., Bogden J.D., Blood lead concentrations and pregnancy outcomes [J], Arch. Environ. Health, 2002, 57(5): 489-495.
[10] Zhang S.N., Liu C.F., Dong J.W., Li X.J., Xiang C.Q., Effect of lead on reproductive endocrine function in pregnant rats [J], Lao Dong Yi Xue, 1995, 12(2): 9-11.
[11] Govoni S., Lucchi L., Battaini F., Spano P.F., Trabucchi M., Chronic lead treatment affects dopaminergic control of prolactin secretion in rat pituitaty [J], Toxicol. Lett., 1984, 20(3): 237-241.
[12] Liu W., Investigation on reproductive function of lead workers in township enterprises [J], Henan J. Prev. Med., 2001, 12(2): 95-96.

Retraction

Ying Chen, Guoling Wang, Jianyong Wu, Qin Xiao, Jiahai Lu

Evidences of Swine Influenza Infections among Swine Workers by One Health Approach

ISBN (Online): 9783110559040
DOI (Chapter): https://doi.org/10.1515/9783110559040-025
DOI (Book): https://doi.org/10.1515/9783110559040

Retraction of: Chen, Ying / Wang, Guoling / Wu, Jianyong / Xiao, Qin / Lu, Jiahai; Evidences of Swine Influenza Infections among Swine Workers by One Health Approach, Proceedings from the ICERP 2016- International Conference on Environmental Research and Public Health, 2017, 229 – 235

The authors have requested to remove their article because much of the text and results was published previously.

Kolibri 360 GmbH
Ferdinand-Jühlke-Straße 7
99095 Erfurt, Deutschland
produktsicherheit@kolibri360.de